The Aviators

The Aviators

by
William Joy

Shakespeare Head Press
Sydney

Shakespeare Head Press.

First Published 1965.

Reprinted 1971.

ISBN 0 85558 011 9

Printed by Toppan Printing Co. Ltd., Hong Kong.

Acknowledgements

The author wishes to acknowledge the help given him by the librarians of the Public and Mitchell Libraries. Much of the information in this book was taken from contemporary newspapers and other journals and records in libraries. Most of the pictures came from the archives of Consolidated Press. The author also received much help from the library of Qantas Empire Airways.

It should be noted that many of the photographs reproduced in this book are very old and hence do not measure up to the technical standards of modern photography.

Introduction

All who study the history of aviation must be impressed by the number of Australians who star as trail blazers of the air. Harry Hawker was the first pilot to challenge the Atlantic and was unlucky not to get across. Ross and Keith Smith won the race to Australia in which some Australians died. Bert Hinkler slashed their time when he made the first solo flight to Australia in a baby plane with a four-cylinder engine and wings that folded so it could be stabled in a garage. The peerless Kingsford Smith and his heroic mate C. T. P. Ulm were the first to conquer the Pacific. Hubert Wilkins made aerial history in the Arctic and Antarctic and blazed the ice trail across the north pole from America to Europe. Harold Gatty, a Tasmanian, was navigator to the one-eyed American Wiley Post when they made a record round-the-world flight — 15,500 miles all in the northern hemisphere.

The total population of Australia in the heroic years of the air — 1920 to 1938 — was from five to seven million, a fraction of the populations of America and Europe. Why then were Australians so numerous and eminent in world aviation?

First was the sense of adventure never far removed from Australians — as exemplified in both world wars when Australians more than held their own in numbers and courage in the air as they did by sea and land.

A second but equally potent factor was the early realisation by Australians that aviation had come to stay, that aircraft, so flimsy and unreliable in their early days, would one day slash the weary travel times that separated their country from Britain and Europe. The sense of distance, a feeling of isolation, had been with Australians ever since the First Fleet took eight months to make the journey out. Fast clippers cut the time to 91 days by 1849 and 63 days in 1854, while steam brought the two countries steadily closer in point of time. The submarine cable and Australia's Overland Telegraph Line reduced verbal communication with the world to minutes instead of weeks, but the sense of distance remained.

Australia's air pioneers realised that aviation would annihilate distance. With unquenchable courage they set to work to prove it, spurred in some cases by the hope of rich rewards for those who were first in the field. Some were to die in the process. Others were to say with Kingsford Smith: "We chaps who blaze air trails have little to show for our deeds. Ill fate has marked us for its own." Through all the turmoil of death, intense effort and vain hopes came success. Today the jets of Australia's own world air line, on special charter flights, sweep from Sydney to London in 24 hours, itself a tribute to the faith and persistence of Australia's air pioneers.

Illustrations

Contents

— Chapter 1

The balloon age

Back in 1850, an eminent medical man of Sydney, Dr. William Bland, surprised his fellow colonists by producing plans for an Atmotic Ship or flying machine which, he said, would carry them from Sydney to London in four or five days. Bland, who had been transported for killing a ship's purser in a duel and quickly pardoned, planned a large gas bag with a passenger or cargo float driven by steam. As with most innovators, Bland's brain child had a mixed reception. Cynics laughed and jeered. The whole idea of flying to England was nonsense, they said. Others, admiring Bland's record in medicine and politics, his research to prevent spontaneous combustion in wool ships and fires in coal mines, were not so sure. There might be something in it, they thought. There *was* something in it. Bland's Atmotic Ship was years ahead of its time. Some now consider it the precursor of the zeppelin.

Bland, however, was a long way down the list of famous men who dreamt of flight. Rameses III of Egypt is said to have rigged himself wings 3000 years ago but baulked at testing them. Simon, the magician, crashed to death from a tower in the Circus Maximus when Roman Emperor Nero ordered him to prove he could ascend heavenwards in a fiery chariot. In 1489, Leonardo da Vinci built a "great bird" in which his assistant, Zoroastro, jumped from a barn and crippled himself. The first successful step to flight came in 1783 when Joseph and Etienne Montgolfier, silk merchants of Annonay, France, invented the hot-air balloon. A Frenchman, Pilatre de Rozier, was the first man to fly in their balloon — and the first to die when he tried to combine hydrogen with fire. When Dr. Bland planned his Atmotic Airship, ballooning was common, though dangerous, in

Europe. Bland, however, was not interested in ballooning as a thrill or a sport. He wanted to fly to England.

Bland's Atmotic Ship, if built, would have resembled a giant sunfish, with a gasbag 200 feet long, 90 feet high and 60 feet broad. As with modern airships, Bland intended to divide the bag into a series of gas-tight compartments to reduce the fear of bursting. Below the bag, he planned 80 feet of studding sails and a car 100 feet long by 30 feet wide, holding a stove, a boiler, a tank of water and a locker for coal, with an asbestos screen to stop the sparks. Bland thought steam thus generated would turn screws, fore and aft, which, after doing their part in propelling the ship would turn windmills fixed to each end and give added impetus. He estimated he would get 40 knots from sails and 10 from steam. Grandly he announced that his ship would carry one and a half tons of passengers or cargo from Sydney to London in four or five days.

Bland sent plans of his airship to Queen Victoria and Emperor Napoleon III of France, and was more than peeved when the President of the United States, Abraham Lincoln, did not use it in the Civil War. To each of these he stressed that with his Atmotic Ship explorers could lay bare the secrets, then veiled, of darkest Africa, South America and Borneo, the frozen wastes of the Arctic and Antarctic and the summits of the highest mountains beyond the reach of man. He shuddered to think what would happen if his invention fell into the hands of barbarians who might use it to overwhelm the civilised portions of the globe. Models of Bland's ship were displayed at London's Crystal Palace and at the Paris Universal Exhibition, but no one would try it.

In 1856, five years after Bland produced his plans for an airship, a French aeronaut, Mons. P. Maigre, arrived in Sydney with a hot-air balloon and promises to thrill the people with the first ascent ever made in Australian air space. His effort to fly ended in a riot in which a boy was fatally injured and much property was destroyed by an angry mob. The ascent, to be made from the Sydney Domain, was widely advertised with the result that 10,000 people turned up, headed by the Governor, Sir William Denison, and his family, the first Premier (S. A. Donaldson), Mayor Christie and Capt.

McLerie, chief of police. Maigre's sponsors charged the affluent 5/- each for seats in the enclosure. The "groundlings" paid 1/- to watch from outside the pale.

Plans went wrong from the start. Aeronaut Maigre suspended his balloon between two 70ft. poles. Under the mouth of the balloon he fixed a stove on which he burned bundles of straw saturated with spirits of wine to produce the hot air which he hoped would send him and the balloon soaring heavenwards. After an hour, the balloon showed no signs of inflation. The hot air had gone cold.

"That thing won't go up," said Governor Denison. "It's too heavy."

"I'll stake my reputation on it," replied Maigre, stepping up the burning.

By now the crowd was getting restive. They wanted more than music from the Cremorne band for their money. They began to grumble — and jeer.

After three hours, the balloon began to swell. The aeronaut's basket was hitched to it and the order given to let go. At that moment the excited mob, determined to get their shilling's worth, broke into the enclosure, brushed the distinguished paying guests unceremoniously aside and formed a cordon round the balloon. They ran beside it as it dragged the basket over the ground for a short distance until it finally collapsed.

A howl of rage went up. The spectators thought they'd been tricked. A crowd of boys pounced on Maigre, tore his cap from his head and buffeted and hustled him. A dozen police rushed to the rescue. Maigre bolted for his life with 1000 angry people after him. He went to earth in the south lodge of the Domain and crouched there while the mob smashed all the windows.

Meanwhile the rest of the crowd went berserk. They would have turned on the Governor himself had not Capt. McLerie and some of the police formed a guard around him and his party and hurried them home. The gallant captain then raced for the horse police.

Behind him in the Domain, the cry went up: "Fire the balloon!" A score of hands seized it and dragged it to the stove. The best part of a cask of spirits of wine helped it burn. Next the mob set fire to the official marquee and, to keep the blaze going, grabbed the hundreds of seats and chairs in the Domain and fed them to the flames. At last all that was left to burn was the two 70-foot poles. Eager, angry hands dug them up and wrenched them down. One, in falling, struck two boys, Thomas Downs, aged 11, and Thomas White. Downs died of shocking head injuries next day. Friends carried the two boys aside while the mob went on with the riot. A fierce struggle raged for the ensign of the 9th Regiment which surmounted one of the poles. The flag was finally secured and carried off in triumph by Private John Cavanagh.

The riot had raged for nearly an hour when Capt. McLerie returned with his horse-police. The newspapers were very severe on the Captain for what they called the brutality of his followers. They wrote of mothers hiding behind trees afraid to search for their missing children. One accused the Captain of slashing an innocent bystander across the face with his whip under the wrongful assumption that his victim had thrown stones at him. They wrote also of the brutal horse-police chasing onlookers as far as Market Street and of a man knocked down and left groaning pitifully in the gutter.

Capt. McLerie was further attacked in letters to editors. One correspondent referred to the "total inefficiency" of the police. "Had they stood their ground at the onset of the uproar, instead of rescuing Maigre from the ducking he so richly deserved, the whole of the property might have been saved and loss of life prevented." The correspondent reproached McLerie for "fleeing" for his set of "infuriated horsemen", who "galloped frantically in all directions". Another called it the "Great Balloon Hoax" and said the sponsors were well known to be first-rate hands at gulling the public. A third wrote that "Governor Sir W. Denison may now take into consideration the propriety of not infringing on public rights by closing the Domain in future for any mountebank exhibition."

Meanwhile an inquest·was held on little Tommy Downs, Australia's first air accident victim, at the Three Tuns Inn, King St. The jury recorded that "Thomas Downs came to his death by the falling of a pole in the Domain which was thrown down by a disappointed and excited crowd out of which it is impossible to single any individuals as ringleaders. We unanimously consider that if any individual is to blame it is Mons. Maigre, the perpetrator of the sham balloon ascent which we consider caused the death of the boy. We wish this to be considered a censure on Mons. Maigre." The unfortunate Thomas Downs was buried at Camperdown under a headstone adorned with the carving of a balloon and recording the tragic details of death.

In fairness to Mons. Maigre it should be added that a French spokesman claimed that the trouble had been caused by four or five larrikins who had entered the balloon without permission and distracted the attention of those whose duty it was to keep the fabric away from the stove and stop it singeing the balloon or burning a hole in it. Such a hole, a metre long, had been burned in it, with the result that the hot air had escaped. Mons. Maigre had spent an hour repairing it but had not sufficient inflammable material left to make the hot air needed for the ascent. This explanation was printed in French which the newspaper did not bother to translate for its readers. Mons. Maigre bowed out with infamy.

Australia had to wait two more years, until February 1858, for its first balloon ascent. The honour went to Melbourne thanks to theatre magnate George Coppin who engaged two experienced aeronauts from England. Mr. G. H. Brown was deputed to make the balloon, named *Australasian,* while Mr. Dean was engaged as an experienced aeronaut or flying man, having made frequent ascents in London. The local press recorded with awe that the balloon was 60 feet high and 40 feet across. More than 500 yards of French material 42 inches wide was employed in its fabrication, and it needed nearly 31,000 cubic feet of gas.

Melbourne's Cremorne Gardens, the banks of the Yarra, the hills, "every coign of vantage", were crowded on February 1, 1858, the day fixed for the ascent. The balloon was partly inflated at Melbourne gas works, then hauled by horse and cart and 30 men to the gardens where inflation was completed at Mr. Coppin's own gasometer. Unfortunately, owing to the great heat and violent breeze, the needle-holes at the seams were greatly stretched and a good deal of gas escaped. Enough remained to lift only one aeronaut. After stressing his claims as designer and maker of the balloon, Mr. Brown gracefully stepped down in favour of Mr. Dean who took his place alone in the basket.

To a roar of cheering from the crowd, who had paid 5/- each to enter Cremorne Gardens, the huge machine slowly left the ground, narrowly missing the gates at the north end. Breathless silence came when it looked as if the machine would crash. Mr. Dean was seen feverishly throwing bags of sand ballast overboard to enable the balloon to rise. Suddenly the machine shot skywards to tumultuous cheers, while the sole passenger waved his cap to the multitude below. Mr. Dean rose to a height of two miles and came down seven miles away near the Cambridge Arms on the Sydney Road. He was only slightly bruised. Cabmen fought over who should take him back to the cheering crowd in the gardens.

Mr. Brown was not so lucky when, a few days later, he made Australia's second balloon ascent. He complained that he was "brutally handled" by the Collingwood crowd who were "waiting for him" when he came down. Altogether it was a bad day for Brown. "After 16 minutes in the air, I was seized with a violent pain in the right ear," he said, "and was forced to open the valve to descend to a lower level. I got into an air current that would have carried me over the bay but, as light was failing, I decided to descend and land on Collingwood Flat. For no reason, as far as I know, I was set upon by the crowd and treated in a most brutal manner. My hair was torn from my head. I was punched, kicked and bruised all over the body and crushed to the point of suffocation besides damaging my balloon." Brown had a kind word for Mr. Needham who "assisted to extract me from the mob of savages".

Aeronauts Brown and Dean took their balloon to Sydney the following December (1858) when 7000 saw them make the first ascent in New South Wales. Governor Denison again graced the event, his arrival being heralded by a salute from four six-pounder guns belonging to the Royal Artillery. With lively memories of the misfortunes of Mons. Maigre, the Governor closely inspected the balloon and concurred with scientists present that the "inflation was quite sufficient". The balloon rose steadily and crossed the harbour to descend near the gully in Neutral Bay. "Of course, there were discontented and unpractical grumblers," commented one newspaper, "but they seemed fewer and less bilious than usual."

Ballooning in Australia continued to be a chancy business. Mr. T. Gale, who made four attempts in Sydney early in 1870, was jeered by some as an impractical visionary until he performed what was hailed as the most daring feat in balloon ascents ever witnessed in Australia. Finding he had not enough buoyancy to lift him from the ground, Gale sprang out, detached the basket with its grappling iron and sand ballast from the bag, tied the end of the netting together, seated himself on the ropes and, thus precariously perched, sailed over the city to the suburb of Glebe where he brought the balloon down. Mr. Gale later made the first balloon ascent in South Australia.

Far more blood chilling was the peril at Melbourne of Fitzroy-born Harry L'Estrange, known as the "Australian Blondin", for his intrepid walks on a tightrope across Sydney's Middle Harbour. L'Estrange settled in Sydney, decided to make a living out of ballooning and acquired a balloon he called *Aurora.* One version says he made it himself; another claims it was made in Paris during the Franco-Prussian War and used for conveying despatches from the beleagured capital to the provinces. The balloon was constructed of cotton coated with a solution of india-rubber and varnish. It was 50 feet high and had a parachute attached to the centre. Whatever its origins, the story goes that L'Estrange could not get it into the air in Sydney. Blaming the poor quality of Sydney gas, he packed up and took *Aurora* to Melbourne.

Melbourne gas must have been better, for on Easter Monday 1879, *Aurora,* full of company gas, shot skyward from the Agricultural Society's ground in Melbourne. Two thousand paid to get a close-up of the ascent. The rest of Melbourne watched outside. L'Estrange, who had stopped a few leaks with plaster at the last moment, took the precaution of "divesting himself of his watch and chain and other valuables" before climbing into the basket and giving the signal to "let her go". L'Estrange was visibly elated by the speed with which he shot upward. He waved his handkerchief to the crowd and threw out hundreds of cards and handbills.

The balloon had reached a height of about one and three-quarter miles when spectators saw a black hole appear in one side of the gas-bag. The top part collapsed inward. While women shrieked and held their hands over their eyes the balloon plunged earthwards. A parachute opened but only slightly checked the descent. Most of the spectators gave L'Estrange up for dead when the balloon was seen to plunge into the reserve around Government House. The crowd rushed to the spot. They found that, by a miracle, the remnants of the balloon had tangled with a tall fir tree which broke the impetus and brought the basket to a stop just as it reached the ground. L'Estrange, unhurt, was enveloped in the folds of the balloon and fighting for air when rescued.

L'Estrange told a thrilling story of his escape. He had been so elated by his success that he had not realised how high he had ascended. The gas expanded, a factor he realised too late. He had just grabbed for the valve when he heard an explosion like the roar of a cannon. He knew then that the balloon had burst and, looking up, he saw it was rent from top to bottom. Government House was below him, appearing no larger than a walnut. "God help me," he said, and set to work throwing out whole bags of sand, not thinking they might injure someone — possibly in Government House — below. "I was descending so rapidly, I could scarcely keep myself in the basket. I had to hold on tight with both hands. I could do nothing to save my life so prepared myself for the shock. The basket with the sand," he went on, "and the wreck of the balloon weighed about a ton. When it was nearing the ground the velocity increased. All I knew of what happened afterwards is that I felt the basket had struck the ground and I was enveloped in the wreck of the balloon. I was released by a stranger and found that the balloon was hanging from a tree."

L'Estrange was a first-class showman. He dashed off and, from a hall that night, told admiring townsmen of his great peril. "Feeling death would be inevitable," he said, "there was nothing to do but commend myself to a Higher Power." He blamed the quality of the Melbourne gas, which, unlike Sydney's, was better than he had bargained for. The experts, however, said it was nothing to do with the quality of the gas. The fault lay with want of experience on the part of the aeronaut. From that moment Harry L'Estrange faded from aeronautical history.

Other thrilling ballooning feats followed. Beautiful Val van Tassel, clad in circus tights and sitting on a trapeze bar, went 5,000 feet up in a hot-air balloon. She hung by her toes from the trapeze over Newcastle and (later) Bondi, then descended gracefully by parachute. She claimed to be the first woman to make either a balloon ascent or a parachute descent in Australia. Soon, tethered balloon ascents were open to all at Wonderland City, Tamarama Bay, south of Bondi. An enterprising showman, Vincent Patrick Taylor, otherwise Captain Penfold, then became famous in Australia and America as an airship master and parachutist.

By now, however, the idea of controlled heaver-than-air flight had come to an enlightened few in Australia. Foremost among them in 1880 was Lawrence Hargrave, a young assistant astronomer at Sydney University. Hargrave experimented, some said crazily, with model flying machines. As a preliminary he studied the movements of living creatures that twisted, turned, moved upwards or downwards, just as he considered a flying machine should do. He amazed the Royal Society in Sydney in 1884 with his trochoided plane theory which, he said, explained equally the pivotal, flapping soaring of a bird, the fin, tail and body movements of a fish and the muscular circular ripples by which the worm made its way through the earth. He thought such knowledge could be applied to flight.

Hargrave revealed that in forming this theory he had studied the movements of leeches, eels, sunfish, porpoises, snapper, alligators, lizards, slugs, caterpillars and jelly fish. He had watched the flight of skylarks, hawks, partridges, ducks, pelicans and albatrosses. He deduced that all these creatures progressed trochoidally through turning pivots or muscular ripples round pivots. He made working models of a worm and a fish and measured their movements as they squirmed and swam. Most of his listeners thought this was a queer way to learn to fly and branded him an eccentric. Hargrave himself recorded wryly that, after one of his lectures, a member drew him aside and warned him solemnly that, unless he stopped this nonsense, he'd finish in a lunatic asylum.

Hargrave, however, pushed on with his experiments. He designed and built more than 50 model flying machines up to 10 feet in length of cane and paper and other light materials, including light-weight metal tubing. Some were propelled by flapping wings like birds, others by screw propellers. One flew 368 feet which many considered a record.

Hargrave powered his models first with rubber bands, then with clockwork and finally with exquisite, miniature compressed-air engines he made from tin in his workshop. In the process he invented in 1889 the first rotary engine in which cylinders rotate with the propeller round a fixed shaft. Hargrave never patented his inventions which is why he did not benefit when Europeans reinvented the rotary engine, exemplified in the Gnome, in time to fly the killer planes of World War I. If Hargrave had joined his rotary engine to an efficient propeller, he could have been the first man to fly. As it was he preferred the flapping machines and made a man-carrying model worked by hand in which he failed to leave the ground.

In 1892, Hargrave turned to the study of air currents and the lifting-power of kites. He designed a soaring kite with curved surfaces — a notable advance — and in 1894, at Stanwell Park, south of Sydney, was lifted 16 feet into the air by four box kites one above the other. "A safe means of ascending in a flying machine is now at the service of any man who cares to try," proclaimed the jubilant Hargrave. He then experimented with a man-carrying glider with which he jumped off a sand hill. A gust of wind turned it over and wrecked it.

Between 1895 and 1903 Hargrave designed and experimented with machines which he hoped would carry men. The first was to be supported by a string of kites as well as by its wing surfaces. The second, an embryo seaplane on *papier mache* floats, incorporated box-kite wings and was fitted in

turn with steam and petrol-driven flappers. For a third
sea-plane, Hargrave made floats of tin. Again it was on the
box-kite principle to be driven by steam. The engine, as with
its predecessors, was a failure. Always Hargrave was beaten
by his difficulty in evolving an efficient power unit.

So it came about that Hargrave had the mortification of
reading, in December 1903, that the American Wright
brothers, Orville and Wilbur, with the aid of the petrol
motor, more speedily and efficiently developed in America
than Australia, had made the first powered flight in a
heavier-than-air machine at Kittyhawk, North Carolina. The
Wrights denied they owed anything to Hargrave though they
knew of his work and admired it. Hargrave had the
satisfaction later of knowing that when Brazilian Santos
Dumont made the first European flight in France in 1906,
Dumont's aeroplane was mainly an arrangement of Hargrave
box kites.

Except for an enlightened few, Hargrave was without
honour in his own country. Scientists of other nations were
also slow to appreciate his worth. They had forgotten that, as
early as 1894, Octave Chanute, the greatest authority on
aviation of his day, had proclaimed that if there was one man
more than another who deserved to fly through the air, that
man was Lawrence Hargrave. Because of this apathy scientists
of several nations rebuffed Hargrave's offer of models and
photographs for adequate display in their museums.
Australians, however, were shocked out of their torpor in
1910 when they heard that the Deutsches Museum in Munich
had snapped up Hargrave's models and meant to display them
in special quarters for anyone to study. Some branded
Hargrave unpatriotic when a few years later Britain was
locked in combat with Germany whose planes were superior
to ours in the early days of the war. The accusation was
unjust, however. The models went to Munich because others
would not have them and in pursuance of Hargrave's
pronounced opinion that such discoveries and theories should
be available to students and engineers the world over.
Hargrave, too, had a child-like faith that the flying machine
would not be used to take life. "It will tend to bring peace
and goodwill to all," he wrote, "and will herald the

downfall of all restrictions to the free intercourse of nations."

The story of Hargrave's attempts to get his models displayed is one of constant frustration. He first offered his models and photographs to the Technological Museum (now the Museum of Applied Arts and Sciences) in Sydney as early as February 18, 1901. He explained that he wished to make them accessible to aeronautical students so they would not have to start at the beginning and go laboriously over the ground he had covered in the past 17 years. He wished to give them the benefit of his labours. All he asked was that they should be housed in well-lighted glass cases and that the photographs should be mounted, glazed and framed. The museum replied that there were not sufficient funds available for the purchase of show cases. They already had a few of his earliest models, and would be happy to oblige if, by judicious pruning, he would care to bring the models up to date in the limited existing show case. This would have meant splitting the collection which now numbered more than 70 models and would have destroyed the continuity of the display.

In 1905 Hargrave offered his models to Sydney University, pointing out that they would save months of labour to aeronautical students. The University Senate regretted they could not afford £300 for show cases. He then approached the Aeronautical Society of Great Britain, whose officers replied somewhat coldly that they "would bear his generous offer in mind in case an opportunity occurred of taking advantage of it". The Commonwealth Government, Melbourne Museum, America's Smithsonian Institute and certain British museums likewise turned down his offer. It is not surprising, therefore, that when the Munich Museum cabled offering to take the models and house them properly, Hargrave allowed 73 flying machines and box kites to go to Germany, then at peace with the world.

After ignoring his offers for eight years, some Australians now wondered if Hargrave's models, photographs and notebooks could be of value. Questions were asked in Parliament. One brought a written reply from Under-Secretary for Justice J. L. Williams that Premier C. G. Wade had approved accommodation for Hargrave's models in Sydney Technical Museum. But by then it was too late.

The Deutsches Museum in Munich was heavily bombed by the Allies during the last war. Fifty seven of Hargrave's models, including all the box kites, were destroyed. In 1960 the museum generously returned 12 of the surviving models to Sydney to be housed — where they always should have been — in suitable cases in the Museum of Applied Arts and Sciences. The Munich Museum kept four models, replicas of which are being made for inclusion in the Sydney collection. The Sydney Museum of Applied Arts and Sciences also possesses Hargrave's notebooks which were found in the Library of the Royal Aeronautical Society in London and returned to Sydney. At last Lawrence Hargrave is getting the recognition denied him in his lifetime. He died in July 1915, two months after his son and fellow-experimeter Geoffrey Lewis Hargrave was killed at Gallipoli. A friend wrote that the bullet that killed his son also killed Lawrence Hargrave.

By then, however, the world was well launched into the air age which began when the Wright's made their first flight in 1903. Six years later it reached Australia and developed into a race to decide whether Australians or British should be the first to fly on wings "down under".

— Chapter 2

The first flights

An intrepid aviator, Colin Defries, climbed into a flimsy Wright biplane on Sydney's Victoria Park racecourse on December 18, 1909, determined to establish a claim to be the first man to make a powered flight in Australia. He sent the machine careering along at 35 miles-an-hour, bounced 20 feet and rose into the air somewhat uncertainly. Pressmen recorded that he had flown 300 yards at a height of two to 15 feet when his hat blew off. Like lesser folk, Colin Defries grabbed for the hat. To do so, he let go the lever controlling the elevating plane and found himself swooping on a row of drainpipes someone had thoughtlessly strewn across the course. Defries cut off the engine, swung the plane to the left and came down in Australia's first forced landing, damaging two wheels and a stay. "It's all my own fault," he said sadly. "I should not have worn my ordinary hat."

Because of this mishap some deny that Colin Defries made the first powered flight in Australia because he obviously did not control his plane when it landed. They award the palm variously to Fred Custance, a young Australian mechanic who, pressmen said, flew three miles near Adelaide on March 17, 1910, or to Harry Houdini, the strait-jacket escapeologist, who flew more than two miles near Melbourne a day later.

By 1909 Australia had become increasingly air conscious. Everyone wanted to see these strange flying machines. Enthusiasts even formed an Aerial League to stage exhibitions and prod the Government into encouraging aviation. The first to see money in it were those enterprising showmen, J. and N. Tait who brought aviator Colin Defries from England to fly a Wright biplane in a grand Flying Fortnight in December 1909.

Simultaneously, a band of Australian enthusiasts led by George Augustine Taylor, *Bulletin* and *Punch* cartoonist, secretary of the Aerial League and a friend and admirer of

Lawrence Hargrave, were working on a glider. Associated with Taylor was a young furniture manufacturer, Edward Hallstrom. Thus developed a race to decide who should be airborne first in Australia.

Colin Defries arrived in Sydney and regretted he could not land the plane in Pitt Street. The plane named *Stella,* was too big. He took it to Victoria Park Racecourse where on December 4, 1909, thousands paid 2/6 to J. and N. Tait, confidently expecting Defries to soar forthwith into the heavens. Defries disappointed them. He showed them how he worked the elevators and rudder. A car then towed him and his plane so many times at speed before the grandstand that the bored crowd yelled to him to "have a go".

This Defries did. He failed to get off the ground on his first run. On his second he scattered the photographers. He was going well on the third and was expected to take to the air at any second, when his plane appeared to sink into the ground. He had run into three or four sleepers left across the track and buckled his wheels. Defries took it philosophically. "The elements were against flight today," he said. "I consider it providential I ran into the sleepers. I am a fatalist. I think if I had not run into them I would have gone up and come down very severely in another way. Everything is for the best." The crowd did not think so.

The mishap, however, was providential for George Augustine Taylor, Edward Hallstrom and their friends who had finished their glider. They took it on December 5 to the sandhills at Narrabeen and flew it first as a kite to make sure it was stable and would support a man. George Taylor then took his position on the lower wing. In a series of dips and curves he glided 98 yards from 3 to 15 feet above the sand. Later he travelled 258 yards and finished in the surf. Mrs. Taylor, Edward Hallstrom and others made glides that day, thus assuring that Australians were the first to be airborne in Australia.

Four days later Defries was reported to have made a brief flight of 115 yards though few could vouch for this, for the crowd had become disillusioned and only about 50 saw him do it. He tried again on December 18 and was well airborne in what some regard as Australia's first genuine flight when he lost his hat and crash landed.

Most Australians were unhappy about the Defries fiasco. They wanted steady, sustained, controlled flights. They got them in March 1910, when Harry Houdini, the magician and escapeologist arrived to perform in Melbourne, bringing with him a second-hand Voisin biplane. Houdini's favourite stunt was to jump heavily manacled off bridges into rivers, divest himself of the shackles underwater and swim happily to the surface to the relief of gaping onlookers who expected him to drown.

Houdini "garaged" his plane in a paddock at Digger's Rest, 20 miles from Melbourne, where he was joined by an Englishman, Ralph C. Banks, who, according to Houdini, had taken charge of Defries' Wright biplane. Ignoring Defries' claims, each announced he meant to be the first man to make a powered flight in Australia. Aeroplanes were flimsy things in those days. The slightest breeze could crash them. The two aviators sat side by side in the paddock waiting for a fine day.

The suspense was too great for Ralph C. Banks. Though Houdini shook his head in warning and motored back to Melbourne, Banks decided to attempt a flight on March 1, 1910. He had his biplane nicely in the air and had flown 300 yards when a gust upset it. It dived into the ground and turned a complete somersault. Banks was thrown from his seat and was seen struggling in the wreckage. He escaped with a black eye, torn lip and a bad shaking.

Houdini waited 17 more days for the right weather. On March 18, 1910, he climbed into the Voisin and took off, trying his wings with a first flight of one minute. On the second flight he flew two miles round the paddock. He misjudged the landing and the plane ran along on its nose wheel for some distance with its tail in the air. Taking off again, he flew more than two miles and made a perfect landing. "I can fly! I can fly!" he exclaimed jubilantly.

Harry Houdini promptly claimed to be the first man to make a powered flight in Australia. He spoke too soon. Within days reports strengthened that a young Australian mechanic, Fred Custance, had flown three miles in a Bleriot monoplane near Adelaide on March 17. Houdini had waited one day too long for his fine day. According to reports in the Adelaide newspaper,*The Register,* Fred C. Custance, who had been associated in setting motor records in Australia, made his flight in a Bleriot monoplane imported by Mr. F. H. Jones of Adelaide, who after attending an aviation meeting at Rheims, France, had become enthusiastic about flying. The Bleriot was crated to Adelaide and assembled in a paddock near Bolivar, 10 miles north, by foreman engineer C. W. Wittber and mechanic Fred C. Custance. The first trials were made on Sunday, March 13, 1910, when Custance is said to have had one trial run while Wittber, in tests, taxied at varying speeds and managed to get the plane 10 feet off the ground in a series of 120-yard hops.

Wittber was not present when Jones drove with Custance to Bolivar in the pre-dawn darkness of March 17. According to *The Register,* Custance took his seat in the Bleriot about 5 a.m. "A few preliminary twists of the propeller," runs the report, "and the machine was under way at a good speed. It rose quickly and, with the fences of the paddock as a guide, the area was covered thrice in rapid succession — a distance of about three miles. The height of flying was between 12 and 15 feet. The machine was in the air about 5 minutes 25 seconds, which constitutes a duration record for Australasia."

After waiting for daylight, continues the report, Mr. Custance, "anxious to eclipse the altitude of 30 feet said to have been attained by Mr. Colin Defries," again entered the plane. The machine started off in wonderful fashion from a 40 yards run and quickly mounted to a height of about 50 feet. After travelling about 200 yards, Mr. Custance made a slight error in manipulating the elevators and caused the machine to descend suddenly, head foremost. The under-carriage was smashed and the propeller broken while Mr. Custance was thrown through the framework and struck his head against the petrol tank. The newspaper enthused that the young mechanic should have achieved such a triumph at his second attempt.

Mr. Jones was equally delighted. He recorded that Custance landed the Bleriot without trouble after his three mile flight. "Custance will make an excellent aerial pilot," he said. "He is keen and has plenty of nerve and pluck. His first flight was like that of an experienced aviator." Custance himself made light of his feat. Flying is not much different from running a plane on the ground, he said, except that you experience no bumps and you have a sense of floating. And as to his crash, the petrol tank was dented, while he had escaped with a few bruises and a headache. Some, over the years, have doubted Custance's claim. They point out that Bleriots were tricky machines and it would be a miracle to fly one in a number of circuits second time up.

Though many quarrelled and still do over who made the first controlled powered flight in Australia, everyone conceded that the air age had at last arrived. A progressive few began to badger the Government to put Australia into the forefront of aerial development. Spearheading the campaign was the first airborne Australian, cartoonist George Augustine Taylor, and Major (later Major-General, Sir) Charles Rosenthal, then commanding No. 5 Howitzer Battery. Taylor and Rosenthal formed the Aerial League in Sydney and proceeded to ginger up the Government. Their first proclamation declared roundly that their aim was to help Great Britain secure supremacy in the air equal to her command of the sea. The proclamation claimed, somewhat extravagantly, that the "modern aeroplane" was Australian in origin, "exploited by others while Australia slept," and that the League would see justice done.

To arouse public enthusiasm, the League staged exhibitions in Sydney's Prince Alfred Park. The star turn was an ascent by Captain Penfold's hot-air balloon. The balloon, however, came down sooner than expected and tangled with the tramway wires. The Town Clerk, T. A. Nesbitt, promptly banned all air displays from Prince Alfred Park. He even sent a policeman to break up the unlawful assembly when George Taylor staged a model aeroplane contest in the Exhibition Building there. Taylor and the young furniture manufacturer, Edward Hallstrom, defied the law. Hallstrom flew his model plane and was pounced on by police. The case was laughed off before it came to court.

Prodded incessantly by George Taylor and his League, the Federal Government offered a prize of £5000 to the first Australian to build an aeroplane in Australia and fly it. The government stipulated, however, that the plane should also hover like the modern helicopter which, in those days, was asking the impossible.

The offer stimulated the Australian imagination. Several mechanically minded men set out to build powered planes, though few knew much about them. George Augustine Taylor was possibly the first to try. He had an engine built at Balmain for a monoplane, found it defective and went to law over it. Messrs. Kolta and Newman built a monoplane with a 20-horsepower engine, took it to Randwick Rifle Range but could not get it into the air. Interest flared again when a New Zealander turned up in Sydney with a model "hovercraft" consisting of a biplane with revolving upper wings. The Tivoli Theatre put it in a variety show.

While others failed, a young bushman, John Duigan, was working on a plane in a sheep station workshop, near Mia Mia, 80 miles from Melbourne. The only things Duigan had to guide him were photographs of existing planes, magazine articles on the American Wright brothers' plane and Hiram Maxim's book *"Natural and Artificial Flight"*. First Duigan made a glider consisting of a pair of wings with a hole in the middle for his body. He crashed. He then copied a Wright brothers' glider and amazed the local sheep men who galloped after him as he flew.

Getting the "feel of the air", Duigan built an air frame, 35 feet long, with 24½ foot wing-span. He used ash and red pine from nearby plantations, wire from a piano and made metal fittings from the steel bands the Duigans used to bale their wool. He would not trust glue, nails or screws. Every part of his machine he bolted firmly into place.

Duigan was stumped for an engine until, by chance, he heard of a Melbourne inventor, J. E. Tilley, who had made a petrol engine but could get no one to take it up on a commercial basis. Ultra-conservative local engineers would not take a chance. "Best leave such new-fangled gadgets to overseas firms who can afford to lose the money," they said. Duigan made a deal with Tilley for both engine and propeller. The engine lacked the power he needed so he himself fitted larger cylinders and a water cooling system, and changed the drive from belt to chain.

The first test was a failure. Duigan said later it was a toss-up whether the machine stood still and the propeller revolved or the propeller stood still and the machine went round it. On July 16, 1910, he had better luck. He got the plane into the air for a hop of 24 feet, "tiptoeing along the ground" as he called it. Three months later Duigan flew his home-made plane 196 yards 12 feet above the ground, at 40 miles an hour, capping it later by flying around Bendigo racecourse to the wild enthusiasm of the crowd.

Meanwhile another colourful character was thrilling the crowds around Sydney. William Ewart Hart, a Parramatta dentist, fell in love with a Bristol biplane brought to Australia by a New Zealander, J. J. Hammond, and Leslie MacDonald, Australia's first aerial barnstormers who in 1911 tried to sell the plane to the Government. The Government would have no truck with such new-fangled notions so Hart was able to buy it for £1300. The Bristol company refused to let MacDonald teach Hart to fly unless Hart promised not to build aeroplanes in Australia. Hart said "No", and proceeded to teach himself: His first plane was wrecked by a storm before he took it off the ground. He salvaged what parts he could and built another.

He had made scores of taxi-ing tests when a friend asked him casually, "When are you going up?" "Tomorrow", replied Hart, jokingly. He arrived at the field near Penrith next day to find 500 people waiting to see him make his first flight. Hart was not the man to back out. He coaxed the machine into the air and by good luck made an easy landing. From that month Hart was constantly in the headlines. His greatest feat was to land on the Agricultural Showground at Moore Park in Sydney. A golfing enthusiast, watching the flimsy plane approach the high barrier of hoardings, shouted in dismay, "He'll have to try a lofting shot over that bunker." According to one report that is exactly what Hart did. An eye-witness stated, "He cleared the barrier with a beautiful rise and dropped on the green as lightly as a bird."

Getting out of the showground was more difficult. Hart had to get spectators to hold his machine while he revved up, got off to a flying start and, once again, just cleared the

Mr. Gale made his balloon ascent in the Domain, Sydney, in January 1870, after several failures. The balloon had not sufficient gas to lift car and man. Gale, therefore, unhitched the car and made the ascent seated on the rope netting.

A contemporary picture of Harry L'Estrange's balloon after it had burst high over Melbourne in 1879. The balloon fabric tangled in a high tree, thus saving L'Estrange from death.

hoardings. He stated later that he did not realise what he had taken on and would never do it again.

Hart, who held Australia's first pilot's licence, was constantly in trouble. He crash-landed in a Chinese market garden and had to compensate the angry Orientals. Later he walked away unhurt when a new monoplane he was testing crashed. He lost height dangerously through overloading while flying Major Rosenthal to Parramatta. Compelled to make a forced landing, he headed for the only clear ground, the railway line near Mount Druitt. The plane struck a signal post and slewed off the line just as the Mountains Express came roaring along.

Hart was soon in trouble with the law. Dairy-farmer Hugh Byrne hauled him before the District Court for flying over his 320 cows and stampeding them. Byrne claimed that two died and the rest went off their milk. Judge Backhouse awarded Byrne £20 damages.

Hart had flown 3000 miles in 500 flights and had carried 150 passengers when an American, Eugene (Wizard) Stone, challenged him to an air race from Sydney to Parramatta, a distance of about 15 miles. Stone had been giving displays in his nippy monoplane, filling in time looping the loop on motor cycles with his wife in the Globe of Death. The race, billed as the First International Aviation Contest, was fixed to start from Ascot Racecourse on June 15, 1912. Fifteen thousand paying spectators packed the grandstand and paddocks. However, the day was too gusty for racing, and the crowd had to be content with displays of crazy flying in which the two aviators tried to outdo each other in daring.

The race came off on June 29, starting from Surrey Park. Hart won the toss and was first away, flying low on course towards Parramatta. Stone followed five minutes later, rising high so that he could keep Hart well in sight and creep up on him, but a rain cloud swept up and obscured his vision. When Stone emerged from it, Hart was nowhere in sight. Cheers rose from the crowd when news came that Hart had landed safely in Parramatta Park, 23 minutes after starting. They had to wait longer before hearing that Stone by some strange mischance had lost his way and had landed near Lakemba. Stone claimed somewhat sourly that he'd been misdirected.

"I was given certain directions before starting but they were not much guide," he said. "I was told to keep the river on my right when I caught sight of it. This I did, but it was the wrong river. I was told I would see a town but the town I took to be Parramatta was another one altogether."

Hart crashed while testing a new monoplane on September 5, 1912, suffering a compound fracture of the left leg, a broken right kneecap and complicated fractures of the skull. For days he hovered between life and death, then slowly recovered.

Another aviation thrill came to Australia in April 1914, when an enterprising Frenchman, 31-year-old Maurice Guillaux, landed in Sydney from the liner *Orontes* with a crated Bleriot monoplane. Lord Mayor Richard Richards gave him an official welcome after which he proved his skill with the Bleriot by flying to Newcastle. Guillaux was lucky. Postal chiefs had already planned Australia's first air mail and had engaged the American "Wizard" Stone to fly the mail from Melbourne to Sydney. Stone crashed his plane before the date fixed for the great flight. Maurice Guillaux stepped into his place.

Guillaux took off from the Royal Agricultural Showground, Melbourne, at 9.12 a.m. on July 16, with 2000 specially-printed postal cards which were sold to the public for one shilling each. His first stop was Seymour where the road was crammed with motors, waggons, carts, horsemen and crowds of pedestrians massed to see their first aircraft. Guillaux dropped in at Wangaratta and, at 12.59, after encountering many air pockets which tossed his machine about, landed at Albury. At Wagga he failed to see the guiding bonfire so came down on the racecourse where a meeting was being held. He took off at once for a nearby paddock.

Guillaux was now about half way on his 582-mile journey. He lost no time and by 3.30 was at Harden. He left quickly for Goulburn, but a head wind and imminent darkness forced him to turn back. He set out for Goulburn again next day, only to be foiled by rain and cloud. Soon after dawn on July 18, Guillaux took off from Harden for the third time, and battled his way through bitter weather "above those cruel mountains" to Goulburn. The next hop took him to a forced landing near Liverpool.

A tail wind sped Guillaux on the final lap to Sydney where he arrived ahead of schedule, being obliged to fly over Manly to pass the time before landing at Sydney's Moore Park. Guillaux's flying time for the 582 miles was nine hours 33 minutes. In Sydney, his feat was widely acclaimed.

In the war, Maurice Guillaux became a captain in the French air force. He was seconded as an instructor to Shrewsbury air base in England where he trained Australian airmen. Guillaux was ready for any daring escapade as he proved when, for a bet, he took a Farman trainer into the air with four men on the wings and two in the cockpit. He crashed to death at Villacoublay in May 1917.

Long before the first mail flight, experts were convinced that planes would figure prominently in the next war. Some Australians were agitating for an Australian Flying Corps. Leading the attack was Aerial League secretary George Augustine Taylor. As early as 1911, Taylor campaigned in the Press for military flying schools, a corps of Australian aerial engineers and for the military use of wireless. He capped the campaign with a withering blast on what would happen if war came. "Australia, at the crash, will be like a rabbit in a trap," he thundered. "It will watch the blow fall and be powerless — for the crash will come from the sky. You, O brother, may rush the fighting line with bare fists if nothing else. But what of your wives and children? Some, at the first thought, will rush the great buildings till they are shattered. Others, more desperate, will crowd and fight and tear at the manholes of our sewer shafts and will mass in huddling mobs in the darkness, safe from the aerial terror that is smashing their menfolk above. And this day will come as sure as tomorrow's sun," he warned, "unless Australia is prepared to meet aerial invasion with aerial defence."

With such rising agitation, the Australian Government just had to act. They established an Australian Flying Corps base at Point Cook, near Melbourne, in 1913. Soon Australians were fighting in the air.

— Chapter 3 —————————————————

War in the air

India launched Australia's first war birds into the air. In February 1915, the Indian Government cabled Australia for pilots and planes to cover the campaign in Mesopotamia where General Sir Charles Townshend, commanding the Sixth Indian Division, was planning a drive to Bagdad. This, combined with the Gallipoli campaign which aimed at Constantinople, was expected, wrongly it transpired, to knock the Turks right out of the war.

Australia jumped at the chance to put fighting men on wings. Within two months the first Australian half-flight was on its way to the Middle East from Point Cook Central Flying School. The half-flight comprised Captain Henry A. Petre, in command, Lieut. G. P. Merz, a medical-man-turned-pilot, Lieut. W. H. Treloar, an Australian trained in England, and military Captain T. W. (later Sir Thomas) White who later became a Cabinet Minister and Australian High Commissioner in London. With the four flying men went 41 other ranks including 18 mechanics. Though they had never worked on aeroplanes before, the mechanics held their own against the best of the British Royal Flying Corps.

The Australians were given crocks to fly. The Maurice Farmans and later Caudrons were totally unfitted for desert warfare. Their top speed was 50 miles an hour and a strong wind would blow them backwards. Their engines were chancy and frequently cut out. A forced landing in the desert meant almost certain death at the hands of marauding Arabs. The only weapons the pilots carried were pistols, rifles and 2 lb. hand bombs which they tossed overboard when they saw enemy troops below.

With such crates, later replaced by Martinsydes which were not much better, the Australians were quickly in strife. Captain White and his observer, Captain. F. C. C. Yeats-Brown, author of *The Lives of a Bengal Lancer,* were forced down by engine failure near Ctesiphon in the Tigris Valley. The engine still ran though not enough to lift the plane. For 15 miles White taxied across the ground running the gauntlet of Arabs and Turks till a bump dislodged the obstruction in a petrol pipe and he was able to take to the air again.

Lieut. G. P. Merz was not so lucky. A dedicated medico, he worked all night with the wounded in the sweltering under-staffed hospital at Nasiriyah. At dawn he took off in a Caudron bound for Basra taking as passenger New Zealand pilot W. W. A. Burn, who preferred to fly rather than make the tedious river trip. They were never seen again.

Captain White, who led a punitive expedition against the Arabs he thought had murdered them, returned with a grim story after burning the house of the Sheik. The Caudron's engine had failed. Merz and Burn had been forced down in hostile country and were attacked by Arabs before they could repair the fault. Armed only with pistols, they broke away in a running fight towards a refilling station 20 miles away. They killed one Arab and wounded five before one of them was hit. The other stood by his comrade. They fought to the end. Both were shot and hacked to death. Their plane was found chopped to matchwood a few days later.

White and Yeats-Brown were next in trouble. Townshend was planning a new offensive against enemy positions on the left bank of the Tigris. They volunteered to fly beyond Bagdad and blow up the telephone wires by which the enemy could order reinforcements from more distant bases. It was a suicide trip over territory teeming with Turks and Arabs. The plane could carry barely enough petrol to take them the 60 miles and back. If a head wind blew up, they wouldn't have a chance.

White and Yeats-Brown set out on Friday, November 13, 1915, taking with them necklaces of guncotton guaranteed to root up the toughest telegraph pole, two revolvers, a rifle, fuses and as many cans of petrol and oil as the flimsy craft would carry. They passed over an enemy force of 2000 cavalry and camelry, picked out a spot that appeared to be free of the enemy and landed about eight miles from Bagdad. A ground breeze caught them and flung the plane against a telegraph pole, smashing part of the lower left wing. Yeats-Brown nipped out with his guncotton necklaces and ran for the next telegraph pole which he blew out of the ground.

According to White's diary, the desert then spawned Arabs. White held them off with the rifle while Yeats-Brown rushed with another charge to blast the wires from the fallen pole and sever them. He was too successful. The wires whipped back and tangled the plane in their coils. The two airmen were helpless, with Arabs converging on them from all sides. The first to reach them was a "hideous black Arab with shaggy hair, stark naked but for two broad bandoliers of cartridges across his chest". The Arabs attacked the two airmen with clubbed rifles and bludgeons. Bleeding and half stunned, they were rescued by Turkish police, who took them in a cart to Bagdad. White subsequently escaped from a Turkish prisoner-of-war camp and made a thrilling journey back to a British base. Treloar also was captured.

Meanwhile, the Mesopotamia campaign came to its disastrous end. Another Australian, Captain Petre, flew out of Kut a few days before Townshend and his 13,000 men surrendered.

By now more and more young Australians were rushing to fight in the air. Some, fretting at the delay, paid their own fares to England to gatecrash the Royal Flying Corps. Others arrived in the Middle East in the Light Horse and quickly changed to the Flying Corps at the call for pilot volunteers.

Among those who served with the British Royal Flying Corps was Charles Kingsford Smith, who first had his baptism of fire as a despatch carrier on Gallipoli. Kingsford Smith had notched several kills when a German scout dived on him from cloud while he was stalking a couple of enemy two-seaters. It was a case of the stalker stalked. Smith had just opened fire on his targets when a burst of machine-gun bullets ripped through his plane. One struck him in the foot, slicing off two toes and, as he put it, a "bit of meat". He blacked out. When he revived he was spinning downwards, his left leg paralysed, with bullets still ploughing into his machine. His flying boot filled with blood past the knee. Despite growing weakness he pulled out of his dive, staggered back to base and made a tolerable landing. Riggers counted 180 bullet holes in the machine, dozens of them within inches of where his head had been.

The King pinned the Military Cross on Smithy at Buckingham Palace. As Smith stepped back his crutches slipped and he fell. The King helped him up, whispering as he did so, "Get out the easiest way," thus waiving all ceremony. Smith turned and hobbled out. "I was the only man there," he said later, "who could turn his back on the King."

Another young Australian who joined up in Britain was Herbert John Louis Hinkler of Bundaberg, Queensland. In his youth, Hinkler had built a glider and toured with aerial stunt man "Wizard" Stone. Hinkler joined the Royal Naval Air Service as an observer and air gunner and made 122 flights over enemy territory, 36 of them bombing raids. He was sick of the carnage when he became a pilot and had to shoot up the enemy as they retreated in the final stages. "This blessed war is making me a cold-blooded murderer," he complained.

Some Australians did not wait for Britain to get going. Walter Oswald Watt of Sydney was so convinced that the French would be first in the air that he joined the flying section of the Foreign Legion. He was forced down in no-man's-land and escaped under heavy fire. He was famous along the whole front when in 1916 he transferred to the Australian Flying Corps in command of No. 2 Squadron.

Other great Australians served in the British Royal Flying Corps. They included Sir Gordon Taylor, G.C., Air Marshal Sir Arthur Longmore and Air Chief Marshal Sir W. G. S. Mitchell.

Meanwhile, Australia was forming her own squadrons. Several served on Europe's Western Front, bombing enemy objectives, shooting their way through dog fights and even tangling with the famous Richthofen Circus. No. 2 Squadron's first big action came in November, 1917, when, still virtually untried, they supported the tanks and infantry of General Byng in his bid to break the German lines near Cambrai. Hunting in couples, 20 or 30 feet from the ground, through patches of thick fog, they flew up and down roads and trenches emptying their machine guns at pockets of German troops and dropping bombs on batteries. One pilot flew into a haystack and survived. Another, Lieutenant H. Taylor was shot down well inside the German lines. He joined an advanced British infantry patrol, led it forward, brought in a wounded man, then returned to his own base.

Major-General Trenchard, commanding the Royal Flying Corps in the field, said the work of the Australians was magnificent. "These pilots came down low and fairly straffed the Hun," he wrote. "Flying among treetops, they revelled in the work. They are splendid."

Aerial battles flared with the Circus and sweep formations. Steadily the British won supremacy in the air. The tally of kills by Australians mounted. Grim rivalry developed among the Australian squadrons as to which would shoot down the greatest number of the enemy. When the war ended, No. 4 Squadron had notched 137 kills and had forced down a further 72. In second place came No. 2 Squadron with 94 kills and 91 forced down, while fifty-three planes had fallen to No. 3 Squadron. The Australians had lost 60 aircraft.

The star of No. 4 Squadron was Captain Arthur Harry Cobby, a former bank clerk, who shot down 32 enemy planes and 13 balloons. In March 1918, Cobby, carrying as mascot an aluminium figure of Charlie Chaplin, downed two of Richthofen's Circus in one dog fight. As the Allies built up their air offensive the Germans, to conserve planes, often kept their pilots grounded. Cobby and his men swept over enemy dromes and taunted them into the air by dropping old boots with challenging messages addressed to "the footsore aerial knights of Germany". Angry Germans who took off to meet the challenge added to his tally of kills.

When the Allies launched the offensive that ended the war Cobby led the fleet of 80 planes that wiped out an enemy drome near Lille. He flew so low that the wheels of his Camel touched the tarmac before he got his plane airborne again. Cobby was an air-commodore in World War II.

Meanwhile other Australian airmen had won enduring glory in the Middle East. While their mates in Europe were shooting down the enemy and helping Australian land forces break the Hindenburg Line, the First Squadron of the Australian Flying Corps had won supremacy in the air over Palestine. They did so well that, when the Turk yielded,

Field-Marshal Allenby proclaimed to the Australians: "You won us absolute mastery. You put victory within our grasp."

Australia's No. 1 Squadron arrived at Suez on April 14, 1916. They were a raw lot then. Pilots and observers from Point Cook Central Flying School had little knowledge of gunnery, armament, photography or bombing. Artisans had received plenty of parade ground drill but little aircraft training, apart from how to swing a propeller. Their sole equipment on landing was two cars and seven motor cycles. Planes and war training came from the British Royal Flying Corps. Other volunteers, some blooded on Gallipoli, changed briskly from Light Horse saddle to cockpit. Soon they were fighting in the air.

At that time the war in the Middle East was going against the Allies. Australian, French and British troops had left Gallipoli. The Mesopotamia campaign had collapsed in surrender at Kut. Egypt was threatened. Rabid Senussi tribesmen raided west of Suez. To the east, the Turks held the Sinai Desert, the Negev wilderness and Jerusalem and the Holy Land beyond. The outlook was grim. The Turks and the Germans who stiffened them had to be beaten back and hammered into submission. The First Australian Flying Squadron and the Australian Light Horse were to play a major role in this.

At first Australian airmen and their British RFC comrades in the Middle East were given outdated two-seater BE2Cs to fly. They were armed with cockpit swivel guns but could not fire through the propeller. They were no match for the speedy German Fokkers and for the Aviatiks which fired straight ahead. "We really had little chance against them," wrote one pilot. "We depended largely on luck." The Australians were handicapped, too, when bombing. "We had to go without observers," said the pilot, "and, though we carried a machine gun, it was quite impossible to fly the machine and use the gun, too."

Despite their inferior planes, the Australian airmen won a strange ascendancy over their German rivals. They reconnoitred the area and bombed the Turkish lines when General Sir Charles Chauvel threw the Australian Light Horse into the battle which tumbled the enemy from Romani. They had some losses bombing vital railways and dumps and photographing enemy territory before the attack on Gaza and Beersheba.

The Australian airmen brought off some amazing rescues, too. Lieutenants P. W. Snell and A. J. Morgan made a snap landing in the desert to rescue Lieut. J. V. Tunbridge who was staggering through hostile country racked by thirst after being shot down. A few days later, March 19, 1917, Capt. R. F. Baillieu and his observer, Lieut. Ross Smith, landed near a British pilot who had been shot down and had burnt his plane. Ross Smith held off enemy troops with his pistol while the shot-down pilot clambered aboard. Bullets ripped through their plane as they took off. For his share in that heroism, Ross Smith was awarded the Military Cross.

This rescue feat was eclipsed next day by Lieut. Frank McNamara, a Caulfield schoolteacher. McNamara's Martinsyde was hit by anti-aircraft fire in a raid on Junction station, 25 miles from Jerusalem. McNamara was wounded in the leg, blacked out and recovered just in time to pull out of a spin. As he did so, he saw another plane, piloted by Captain D. W. Rutherford, of Rockhampton, plunge earthwards to land near a Turkish camp. Rutherford was struggling desperately to get his engine started, when McNamara landed his Martinsyde close by.

Turks were rushing to capture pilot and plane. Rutherford clambered into McNamara's machine which started to taxi shakily. McNamara's wounded leg slumped on the rudder bar, causing the plane to swerve and pile up. The two airmen struggled from the wreck, McNamara fired the tangled fuselage, then ran for Rutherford's machine. It was a race between airmen and Turks and the airmen won despite McNamara's dragging leg. McNamara hauled himself to the controls. Rutherford, hoping against hope, swung on the propeller. The chance in a thousand came off. The engine spluttered into life. Rutherford scrambled aboard. In a flurry of bullets, McNamara got the crippled plane into the air and, despite the throbbing agony in his leg, flew it 70 miles to base. For that, McNamara received the Victoria Cross.

Fighting in the air stiffened as Germans and Turks clung stubbornly to Gaza and Beersheba. By that time, however, better planes were dribbling through to Australian and British airmen. Venomous Bristol fighters were more than a match for anything the Germans had. They were a lethal escort, too, for clumsier British planes — which explains why, when Allenby tumbled the Turks from Gaza and Beersheba, planes of the First Australian Squadron drove the fleeing enemy in panic from Majdal ahead of the advancing Australian cavalry. They pursued the enemy northward till, at last, he turned at bay in the highlands before Nablus.

Building up rapidly with Bristol fighters, the Australians became the crack flying squadron of the Middle East. From then on they constantly harried the enemy. They bombed his bases on the Trans-Jordan railway, shot up grain boats on the Dead Sea, set fire to ripening crops as Arab harvesters struggled to get them in, and photographed enemy territory.

Allenby now was planning the final blow. For it, he needed absolute mastery of the air. The Australian First Squadron gave it to him. In two months they shot down 15 of the enemy, drove 27 down and strafed them to wreckage on the ground. The Germans in desperation sent new machines to Palestine. Two pairs of Australians, Peters and Traill, and McGinness and Fletcher in two Bristol fighters took on seven enemy planes in one hectic dogfight and shot down four. Ross Smith and Mustard, Tonkin and Camm, McGinness and Fysh, Paul and Weir are but a few of the Australians who notched kills in the battles that shot the enemy from the sky.

German airmen became so apprehensive that they dared not fly lower than 15,000 feet. The story is best told in diminishing German raids. In one week in June, 1918, German raiders crossed the British lines 100 times. In August only 18 ran the gauntlet of Australian fliers. In September the number dropped to four.

In the meantime Australian airmen and their British comrades stepped up the blitz on railway depots, camps and troop formations from Nablus east to Amman and El Kutrani on the Hejaz Railway. In a special mission, Ross Smith flew Col. T. E. Lawrence — Lawrence of Arabia — from General Allenby's H.Q. to a secret rendezvous in the desert. There Lawrence rejoined his ragged army of European adventurers

and friendly Arabs, who proceeded to blast great gaps in the
Hejaz railway and to impose a reign of terror on the enemy.

Meanwhile, protected by the crack British and Australian
squadron, Allenby prepared the great bluff that ended the
war in the Middle East. He built a huge dummy force by the
Jordan. A cock-a-hoop Anzac mounted corps helped make
15,000 "horses" of scrap iron and blankets. Horse-drawn
sleds raced across the desert raising dust and leaving tracks as
though a great transport corps were on the move. The few
German observation planes, forced high by the Australian
squadron, saw the clouds of dust and reported to German
commander von Sanders that Allenby was about to launch a
great offensive on the Jordan. Von Sanders moved his most
experienced troops to that sector to meet the non-existent
threat while Allenby massed his armies near the coast where
the enemy was weakest.

The first blow in the Battle of Nablus, or Armageddon,
was struck by the Australian Ross Smith, now flying a big
Handley Page bomber. With Lieut. E.A. Mulford, Lieut. M.
D. Lees and Lieut. A.V. McCann, he took off at 1.15 a.m. on
September 19, 1918, and smashed the enemy's central
telephone exchange at El Afule with sixteen 112 lb. bombs.
Three hours later the artillery bombardment began. Five
infantry divisions broke the Turkish front by the coast. By
7.30 a.m. the Australian Light Horse and other cavalry were
pouring through the gap, swinging in a great arc to the north-
east to trap the Turkish armies.

Before them flew the planes of the First Australian
Squadron, bombing, strafing, watching the whole enemy
front collapse. Australian Captain A. R. Brown and Lieut. A.
S. Nunan spotted the whole Turkish 7th Army retreating
down a precipice road on the Wadi Fara, a straggling column
of horses, waggons, guns, infantry and cavalry. They scored
five hits on the transport and raked the rest of the struggling
mass with machine-gun bullets. From then it was sheer
massacre. The Australian squadron and a British squadron
dropped six tons of bombs on the army on the precipice and
sent 44,000 machine-gun bullets ripping into it. The gorge
was choked with burning lorries and wrecked guns.
Maddened horses and camels dragged men, guns and

transport over the cliff. British mopping-up troops found 87 wrecked guns in that wadi, 55 wrecked motor lorries, four staff cars, 75 carts, 637 four-wheeled wagons and scores of water-carts and field kitchens.

Pilots of the Australian squadron were flying overhead on October 1 when Australian cavalry entered Damascus to ɪɪop up the last-ditch Germans. The war was over. To some of the adventurous young Australian airmen in Europe and the Middle East there was only one way to get home. That was to fly.

─── Chapter 4 ───────────────

The Hawker miracle

April 25, 1919, was a big day for Australian servicemen in London. The blood, mud and desert war was over. The enemy had been hammered into submission in the air and by land and sea, and the Australians had played a vital role in smashing him. It was right and proper, therefore, that they should have their own grand victory march through London and that it should be staged on April 25, the anniversary of the day Australians stormed ashore on the beaches of Gallipoli to found the tradition of Anzac. The handsome young Edward, Prince of Wales, was to take the salute from a dais outside Australia House in the Strand, in company with the lordly and aloof Field-Marshal Earl Haig, plain "Doug" to Australians, General Sir William (Birdie) Birdwood, and a restless little dynamo W. M. (Billy) Hughes, Prime Minister of Australia.

Australian land forces were jubilant, particularly the infantry, who were to march through London with fixed bayonets, a privilege rigorously denied the general run of troops. The only regrets came from Australian war pilots, grounded, bereft of planes, kicking their heels on leave in London waiting to be demobbed. They could have marched with the rest, but somehow they thought they should fly. Spirits rose briskly, therefore, when word went round that certain planes from Handley Pages to Camels and Sopwith Pups would be available at aerodromes round London on the day of the march. Australian airmen made beelines for them.

The sky began to fill with planes as General Sir John Monash, astride his famous grey, led a long column of Australian cavalry, artillery and infantry toward the saluting base before Australia House on the great day. This was no conventional fly-past. This was a victory show and the Australian pilots were out to make the most of it. One of the boldest, Capt. A. H. Cobby, recorded that 50 or 60 machines were jockeying round the sky "in a mad scramble to pass above the Prince". They looped, rolled and spun ever lower till spectators on the roofs cowered. "Some of us went down into the Strand to do the thing properly," wrote Cobby. "I had to fly for about three-quarters of a mile before I could get out. Finally I was able to zoom up just short of Trafalgar Square." He decided, looking back, that "it was probably the most foolish thing I ever did".

Some critics were severe about these escapades and the behaviour of those Australians who, after the march, tried to souvenir the dais on which the Prince had stood — in pieces. London, however, was tolerant that day. As *The Times* put it, "The Australian soldier has a way of pleasantly and pertinaciously doing just what he chooses and of persuading the London public, and even the London police, that because he is doing it just to please himself, there can be no possible harm in it."

Now the danger — and the fun — were over. Australia's fighting men had to get home, and, to some young war pilots, the only way was to fly. No one had flown from England to Australia before. No one had surveyed a complete route. That did not worry them. Give them any old plane from Government surplus and they'd get there some-how. Some sought wealthy sponsors. Others badgered poor Prime Minister Billy Hughes who had visions of a route east littered with the wrecked planes and broken bodies of Australia's air heroes.

To bring it all under control, Hughes announced in March 1919, that the Commonwealth would give a prize of £10,000 to the first Australian crew to reach Australia in 30 days before the end of the year. The pilots were elated. This was just what they wanted. Some Australian citizens, however, damned the project as a waste of public money. Others said bluntly that it could not be done.

The Australian airmen were convinced it could be done. Their morale, always high, had just been boosted by the exploits of young Harry Hawker, a blacksmith's son from South Brighton, now Moorabbin, Victoria. Hawker had had a crack at being the first to fly the Atlantic, had been forced down and given up for lost, only to be brought home in a ship without wireless to the hysterical joy of all Britain.

Harry Hawker was one of those adventurous young Australians who saved up their fares to London in the box-kite days with the sole purpose of learning to fly. After some vicissitudes, he found a job as mechanic with the pioneer aircraft manufacturer T. O. M. Sopwith and became his top test pilot. Hawker set many aviation speed and altitude records. He won international fame when he and fellow-Australian H. A. Kauper attempted a round-Britain flight, only to be forced down in the sea off Dublin after completing three-quarters of the course. Early in 1914, Harry Hawker visited Australia with a Sopwith Tabloid plane in which he carried joy-riders at £20 a trip, among them the Governor-General, Lord Denman. Hawker volunteered for active service during the war. The British Government turned him down flat. He was far too valuable as a test pilot and that was a perilous job, too.

Hawker, therefore, was famous among airmen when, soon after the Armistice, the London Daily Mail offered a prize of £10,000 to the first aviator of any nationality to fly the Atlantic. Hawker teamed with Navy Commander K. Mackenzie Grieve to win that prize. They set off from St. Johns, Newfoundland, in a Sopwith plane on May 18, 1919, and headed out over the Atlantic. They were seen to drop their undercarriage, as planned, to lighten the aircraft. Then the whole world settled down to wait for news of them.

Most anxious was Hawker's wife, whom he first had met standing disconsolately in a country lane beside her broken-down car, which he speedily put right. Now she was nursing a two-month-old baby. Muriel Hawker was wildly delighted next day when a message came that Harry had landed in the sea 40 miles from the mouth of the Shannon and was triumphant and safe. All night she and her brother answered telephone calls from friends saying how glad they were. The shock came in the morning when they opened their newspaper. *"Hawker Missing, False Report of Fall in Sea"* ran the headlines. The message had been a mistake or a cruel hoax. A deep silence brooded over the Atlantic.

Most Britons gave Hawker and his companion up for dead.

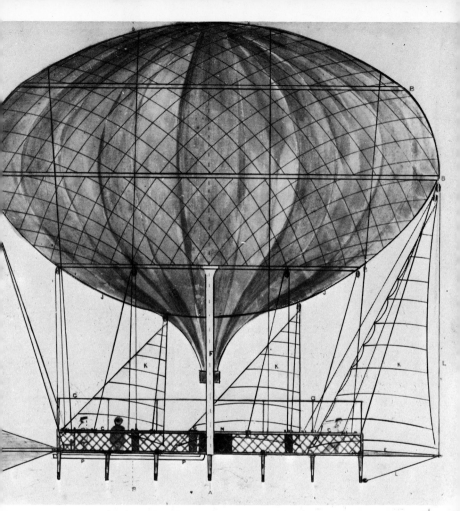

...e front elevation of the atmotic ship designed in 1851 by Dr. William Bland, duellist and
...stralia's first private medical practitioner. Dr. Bland thought he could carry 1½ tons of
...ssengers or freight from Sydney to London in four and a half days in his atmotic ship. Models
...his airship are said to have been displayed in London and Paris. Key to the plan is: A, balloon;
...divisions of balloon; C, car; D, rudder; E, bowsprit; F, sling; G, awning; H, halyards; I, iron-
...od feet; J, mezzialtine; K, sails; L, bowsprit and rudder halyards; M, booms; N, engine room;
...screw.

(Above) Lawrence Hargrave experimenting with box-kites in Stanwell Park, near Sydney, in th [1890s, when he was lifted 16 feet into the air with four box-kites flying one above the other. His box-kite principles were applied by Santos Dumont to the plane in which Dumont made th first flight in Europe in 1906.

(Below) George Augustine Taylor making the first flight ever in Australia at Narrabeen, near Sydney, on December 5, 1909. He and an enthusiastic band of friends built the glider to Taylor's design. A few days later, Colin Defries made a short flight in a Wright biplane. Many, however, reject Defries' claim to be the first to fly a powered craft in Australia.

Some gave three days as the limit in which they might hope; others a week. One of the few who did not despair was his wife. The days dragged by; the world became more pessimistic. On the sixth day, Muriel Hawker received a telegram: "The King, fearing the worst must now be realised regarding the fate of your husband, wishes to express his deep sympathy and that of the Queen in your sudden and tragic sorrow."

On the seventh day, Sunday, Muriel Hawker went to church where she joined in prayers for the safety of her husband. She recorded later, she was confident he would return. About that hour a small Danish tramp steamer, the *Mary*, which had no wireless, was approaching the Butt of Lewis, in the Outer Hebrides. Coastguards saw signals break from her mast. "Saved Hands Sopwith," they read. "Is it Hawker?" signalled the coastguards. "Yes," came the reply. Within minutes a newspaper reporter was on the telephone telling Muriel Hawker her husband was safe. Her prayers had been answered.

The airmen had had a nightmare journey over the Atlantic, battling half gales with a blocked and boiling radiator. At one stage their engine cut out. They were only ten feet above the sea, certain of ditching, when it roared into life again and snatched them from almost certain death. Half way across, they had little hope of making land. They knew they could not keep up much longer. By a miracle they found the Danish tramp steamer, landed in the sea ahead of her and, after two perilous hours, were safely on board. The whole nation rejoiced at the news. The Navy sent a destroyer to pick up Hawker and Grieve and take them to Scapa Flow and then to Thurso where they had the uncommon experience of reading their obituary notices in the newspapers.

From then it was one long triumph for the Australian airman and his comrade. Cheering crowds met them at every station. Inverness, Perth, Edinburgh, and Newcastle greeted them with a din of rattles, hooters, whistles and civic receptions. The King sent a telegram commanding Hawker to Buckingham Palace and another to Mrs. Hawker: "The King rejoices with you and the nation on the rescue of your husband and trusts he may long be spared to you." Muriel

Hawker met her husband at Grantham. He almost fell into the room where she was waiting, she recorded. "He just said the sweetest and most wonderful thing I could ever hear, and added, just as the people started to crush in, 'Don't cry'."

London's greeting excelled the rest. Hundreds of Australian soldiers still there battled to the front of the huge crowd that blocked the approaches to Kings Cross Station. They brushed aside the dignitaries who wanted to give Hawker a civic welcome and caused much concern among the police, who feared they would souvenir the mace. The Australians carried Hawker and Grieve shoulder high through the mob to their car. Some clambered into the car with them, some hitched themselves to it to drag them in triumph through the city. This did not satisfy others. They lifted the car shoulder high. This was too much for Hawker who scrambled over the heads and shoulders of the crowd to a mounted policeman, who carried him tandem on his horse. Later the policeman dismounted. Harry Hawker arrived at the Aero Club on horseback.

King George summoned the airmen to Buckingham Palace, where he stripped away the red tape to give them the Air Force Cross, till then exclusively a Service decoration. A month later Arthur Whitten-Brown and John Alcock made the first non-stop flight across the Atlantic. Two years later Harry Hawker was to die while testing a plane in which he hoped to win the Aerial Derby. Explosions tossed his broken body 200 yards from the crash. The coroner was told he suffered from a tubercular spine and that he probably lost control of the aeroplane through his physical disability. "He was not strong enough to fly and take risks like that." Thus died a gallant Australian who would not give in.

The crossing of the Atlantic naturally whipped up the enthusiasm of Australians, who were determined to fly home, win the £10,000 prize and start the air age in Australia. They chafed at delays and frustrations. Some said harsh things of Prime Minister Billy Hughes and the conditions and rules he sought to clamp on them. In the end seven planes took part in the race. Only two got through. Death lay in wait for some.

Chapter 5

The first race

Australia House, London, became "black with aviators anxious to fly home" when Prime Minister Billy Hughes announced in 1919 that his Government would give £10,000 to the crew of the first plane to fly to Australia. The main stipulations were that the airmen must be Australian, the planes British and that the flight must be completed in 30 days before the end of the year. The recklessness of some of the young airmen appalled Prime Minister Hughes. "One gentleman's plane had a range of 250 miles," he said. "He might possibly have reached Italy. Many of them had no idea of navigation or even of the geographical sequence of the route." When he asked them about landing places and petrol supplies, they waved their hands and brought out a letter from some oil company wishing them well.

Billy Hughes felt respo. sible for the lives of these young heroes who were so eager to fly over uncharted seas and jungles. He ordered strict rules to be clamped on the contest. Airmen had to enter through the Royal Aero Club which would vet them for flying ability and navigational knowledge. The first to fail in this rigid test was young Gallipoli veteran and war pilot Charles Kingsford Smith who was all set with Queenslander Val Rendle and Cyril Maddocks to fly a Blackburn Kangaroo in the great race when the ban fell. None of them had adequate navigational experience, ran the edict. Kingsford Smith and his mates stormed to an interview with Billy Hughes, bullying and begging at the top of their voices into the ear trumpet with which he exasperated them. General Sir John Monash himself asked Billy to relent but he would not. The answer was "NO".

Another who fell out of the race was Bert Hinkler. Official delays caused Sopwith to sell the Dove they had "given" him to fly. Hinkler thought the delays a deliberate "wangle", possibly to favour certain competitors. "Take it from me," he wrote home angrily, "Hughes, with his big; wide, Welsh, noisy mouth is feeding you the choicest sort of claptrap. He has already got more than his share of advertisement out of the belated flight."

Meanwhile, back in Australia, W. H. (now Sir Hudson) Fysh of Tasmania and Paul McGinness, of Victoria, who together fought the Germans over Palestine, were seeking a backer for the big race. They found him in a wealthy grazier, but before they could complete the arrangements the grazier died. The Australian Government asked the disappointed airmen to survey landing grounds in Queensland and the Northern Territory for competitors who might reach Australia. From this survey, through much battling, grew the great Qantas Empire Airways.

The race to be the first to fly to Australia began on October 21, 1919. First away was Scottish-born Captain G. C. Matthews, ex-Light Horse, ex-No. 4 Australian Flying Squadron. Matthews was in a great hurry. Accompanied by Sergeant T. D. Kay, of Ballarat, he took off in a Sopwith Wallaby determined to catch a Frenchman, Etienne Poulet, who had stolen a march on them. Though not officially in the race, Poulet had left a Paris airport a week earlier, determined to win for France the honour of being the first to fly from Europe to Australia.

Matthews was unlucky. Snow and rain bogged him in Europe. In December, well out of the money, he came down 100 miles from Belgrade. Wild Yugoslavians arrested him and Kay, and bundled them into a cell 10 feet square. After four days on black bread and pig's fat the Australians broke jail, ran for their machine and got safely away. With no chance of the prize, Matthews and Kay battled on. At Constantinople they plugged a leak in a water jacket with chewing gum and powdered asbestos. They flew on a faulty engine through sand-storms and rain as far as Bali, where they smashed their wings in a forced landing.

Meanwhile, other competitors were on the way. Next to start was Captain Ross Macpherson Smith of Adelaide, ex-Light Horse, crack pilot of No. 1 Flying Squadron, who had flown Lawrence of Arabia to a secret rendezvous with Arab wreckers in the desert. With Ross Smith, then 26, was his elder brother, Keith, ex-RAF, and Sergeants J. M. Bennett (Melbourne) and W. H. Shiers (Adelaide), who had serviced Ross Smith's bomber in Palestine. Ross Smith was favourite for the race. With Brigadier Borton he had already pioneered an air route to India in his Handley Page. In Calcutta, he and the brigadier had chartered a steamer to survey a route to Australia. The steamer blew up and they were lucky to escape with their lives. They chartered another boat and went as far as the East Indies where the Dutch Governor-General agreed to lay out an airstrip at Bima to halve the last 1750-mile lap of the proposed air route.

Ross Smith, therefore, knew most of the way and had organised supplies along it when on November 12, 1919, he and his crew set off from Hounslow Aerodrome in a big Vickers Vimy for Australia. The weather was foul and branded officially unfit for flying. Sleet and snow deluged them in their open cockpit. It clotted their goggles and windscreen and covered their faces in a mushy, semi-frozen mask. Their sandwiches froze hard. Wrote Ross Smith in his diary that first day: "This sort of flying is rotten. The cold is hell. I am a silly ass for ever embarking on the flight." They warmed themselves at Lyons, Central France. They were in a hurry, too, for Poulet by now was clear of Egypt and speeding on to India.

Early next morning, November 13, Lieut. R. Douglas, of Charters Towers, and Lieut. J. S. L. Ross, of Moruya, N.S.W., took off in pursuit from snow-carpeted Hounslow. They were flying an Alliance plane "equipped with all possible comfort, including leather-upholstered armchairs." The plane rose sluggishly, overloaded, said some, with emergency foodstuffs. They had flown only six miles when the machine nose-dived into an orchard and turned over. Three explosions tore the wreckage. The airmen died instantly. The race had claimed its first victims.

Capt. G. H. (later Sir Hubert) Wilkins, of Adelaide, then 31 and straight from war photography, was the next away in a Blackburn Kangaroo twin-engined biplane on November 21. With Wilkins were Valdemar (Val) Rendle (pilot), of Brisbane, Lieut. Reg Williams, of Victoria, and Garnsey (Gar) Potts, of Richmond, N.S.W. They battled against foul weather across Europe and were well over the Mediterranean between Crete and Egypt when an oil crank case broke and cut their port engine. They struggled back to Crete on one engine and were plunging earthwards almost out of control when a water jacket blew into pieces and ripped through the fuselage. By a miracle they avoided crashing on Canea and finished up in a six-foot ditch. They were out of the race.

Death also lay in wait for Captain C. E. Howell, of Melbourne, and air mechanic G. H. Fraser, who set out in a Martinsyde on December 4. They left Taranto for Athens on December 9. No more was heard of them till December 20 when news came that a machine had landed in the sea near Corfu where a pathetic sodden bundle of good luck telegrams had been picked up on the beach. Peasants said they heard shouts of distress but could do nothing because of the weather. They lit signal fires but no one struggled ashore. When, at last, fishing and naval boats arrived they found no trace of the airmen. Howell's body was washed ashore a fortnight later. His plane was found in 10 feet of water. Experts deduced that Howell had lost his way and probably see-sawed backwards and forwards across the course for hour after hour seeking a landmark before he came down in the storm-lashed sea.

By now, of the Australians, only the great Vickers Vimy was still in the race. Far ahead was Etienne Poulet in his nippy little Caudron two-seater, intent on beating the Australians to Australia. Ross and Keith Smith and their crew settled down grimly to overhaul him. Buffeted over Europe by fearful winter storms, their wings coated often with ice, they sped on to Cairo to find that Poulet was then at Karachi, 2000 miles ahead. When they reached Karachi he was at Delhi, a day's flight away. Bad luck had dogged the gallant Frenchman. He had made four forced landings. His mechanic went down with raging fever. A tropical storm blew his frail plane back over its tracks. They overtook Poulet at Akyab where the big Vickers Vimy towered over the little Caudron like an eagle over a sparrow. Next day they beat him into Rangoon. The end of the road was there for Poulet. A cracked piston forced him out of the race.

Now only the Vickers Vimy remained. All Ross and Keith

Smith had to do was to arrive in Darwin within the 30-day limit and the £10,000 was theirs. With victory almost within grasp, it seemed that fate lined up against them. Storms forced them to land on an emergency air-strip at Singora, near the Siam-Malay border, where they found that the coolies who cleared the trees had left most of the stumps. By a miracle they missed all stumps but one which tore off their tail skid. They made a new skid on a rice-mill lathe hitched to a shafting hand-turned by four coolies. The commandant drove 200 convicts from the local jail to clear away the rest of the stumps before they could take off.

With only four days to go to win the prize they were in trouble again at Sourabaya. The airstrip there had been reclaimed from the sea. The top crust had set hard over a deep layer of liquid mud. The wheels of the Vimy broke through the crust and sank to the axles. The airmen almost abandoned hope. Here they were stuck in a quagmire, only 1200 miles from Australia. "It seemed as if victory would be snatched from us," wrote Ross Smith. The harbour engineer came to their rescue. At his call natives streamed from every direction bearing sheets of bamboo matting, "literally carrying their homes on their backs". They laid the matting in a roadway 300 yards long and 40 wide over the mire. Two hundred coolies hauled the plane on to the matting. "I could never understand how it stood such rough treatment," wrote Ross Smith later.

Soon they were on the last lap. They spotted the cruiser *Sydney* patrolling far out in the Timor Sea and landed at Darwin on December 10, 1919, to be welcomed by a milling crowd at the local jail. They had won the prize with two days to spare. In Australia, they ran into more trouble. They had to get a new propeller and their engines needed a complete overhaul. Thus it was February 14, before Ross and Keith Smith, now knighted, and their gallant mechanics, Sergeants Bennett and Shiers, flew to Sydney in the big Vickers Vimy.

Sydney went wild with enthusiasm over "these young men who, like Columbus, had pioneered a new track across the world". Roof tops and hills were black with people, as the Vimy appeared, escorted by Flight-Lieut. Le Grice and Flight Commander Leslie Holden, flying Curtis planes, and

Flight-Commander Nigel Love in an Avro. Ships' sirens, factory whistles and a storm of cheering rang over the harbour. Men of the Australian Flying Corps and former Light Horsemen paraded in uniform under General Fox and Colonel A. J. A. Travers at Mascot, where a huge crowd gathered to see the airmen land.

The wheels touched the ground to "thunderous cheers" and the strains of "Australia Will Be There" from khaki-clad bandsmen. Pandemonium broke loose. The airmen had just time to greet their parents before Light Horse admirers pounced on Ross Smith and bore him on their shoulders through the crowd. He was helpless. Once he disappeared from view but was quickly up again. Shoulder high they rushed all four airmen into the official tent where the mayor and aldermen of Mascot were waiting. Such a crowd jostled in and around the tent that it began to sway dangerously. Abandoning the tempting food, dignitaries and airmen bolted to waiting cars which took them to a formal reception at Sydney Town Hall.

Melbourne was less lucky. Fifty thousand people waited at Flemington on February 24 to greet the airmen, only to learn they had been held up by three forced landings. Fewer than 40 were present to see them land at Laverton next day. Melbourne made up for it, however, with a reception at Parliament House and an official luncheon at which Billy Hughes gave them a cheque for the £10,000, which they divided equally among the four. Thus ended the first great air race to Australia.

Everyone mourned two years later when Sir Ross Smith and Bennett, promoted Lieutenant, were killed in England testing a plane in which they planned to circle the world. Keith Smith, who was to fly with them, arrived at the aerodrome just in time to see the plane go into a spin and nose-dive into the ground. He recorded later that at their last meeting Sir Ross had told him: "If anything goes wrong during the world flight, there shall be no complaints, no blame shall be cast on anyone — Australians never squeal."

Meanwhile, as Ross and Keith Smith were being welcomed home in triumph, two other airmen, Ray Parer and J. C.

McIntosh, were flying stubbornly towards Australia. Theirs was an epic, too.

— Chapter 6

Whisky for Billy Hughes

Back in December, 1919, a Scottish millionaire distiller, Peter Dawson, handed a bottle of whisky to two young Australian airmen and asked them to deliver it to his friend, W. M. (Billy) Hughes, Prime Minister of Australia. The young airmen delivered that whisky eight months later. In the meantime it had been scorched by fire, narrowly escaped dropping into the crater of Vesuvius and had been in grave danger of capture by Arabs.

In delivering the whisky, the two airmen, Lieutenants Raymond J. P. Parer and J. C. McIntosh wrote a grand chapter in the history of Australian aviation. They flew across half the world in a jalopy they had picked up cheap at an R.A.F. disposal dump, setting records for single-engined planes on the way. The second crew to fly to Australia, they used up five propellers and when at last they arrived their plane was a flying wreck. As Governor Sir Walter Davidson told them: "To carry on when each day brought trouble and to win through on your crazy craft meant stout hearts and dogged pluck. Well done, lads." Though they were well out of the money in the big race, Prime Minister Hughes and his Government gave them £1000 as a consolation prize. "Why not another £9000, Billy?" cried a voice at the presentation.

Parer and McIntosh had been among the young adventurers who decided to compete when Hughes' Government offered £10,000 to the first Australian airmen to fly a British plane to Australia. A pocket Hercules, Melbourne-born Ray Parer, had made successful gliders of ti-tree and calico long before he entered the Australian Flying Corps as a mechanic. He won a commission and was a test pilot during the war.

McIntosh got the wanderlust while training to be an architect in Scotland. He migrated to Perth, wandered round for a time and finally became an expert axemen among

Western Australia's high timbers. He enlisted and served on Gallipoli and in France. The War Office jumped at an aerial bomb he invented. When they asked him what he wanted in payment, he said, "Put me in the Air Force." He had just completed his training when the war ended.

Parer and McIntosh badgered all the big aircraft companies for a plane to fly in the big race. Even when the first competitors left they did not give up hope. They had been ordered to stand by for repatriation when a friend told them Peter Dawson, the Scottish distiller, might help them. McIntosh made the first contact. Dawson liked this fellow Scot. He asked McIntosh if he and Parer had enough flying knowledge for such an enormous job and if the plane they had in view was good enough. Then he came down with a cheque and an offer to finance supplies along the route.

The jubilant young airmen went to the disposal dump and picked out a single-engined D.H.9. They worked against time to bring it into trim. Then they ran into trouble. First, to be in the race, they had to convince the authorities that expert axeman and Gallipoli veteran McIntosh was an Australian. The British Air Ministry then decided the plane was not up to standard and advised the Australian authorities to stop them. Australia House passed the problem to Melbourne and said Parer and McIntosh were not to leave till permission came from Australia. When news came that Ross and Keith Smith had landed at Darwin, they were told that the race was over and there was no sense in starting anyhow.

By now Parer and McIntosh were furious with frustration. The end of the year passed and still there was no word from Australia. According to the rules, they were out of the race. This made the angry young airmen more determined than ever. They decided to jump the gun. Permit or no permit, they were flying. They dodged a couple of telegrams sent to stop them, and on January 8, 1920, took off from Hounslow in fog. The fog was worse over France. They made a forced landing and buckled a wheel. They obtained another wheel from Paris where their stained uniforms, Sam Brownes and big Colt revolvers, caused a sensation at the Folies Bergeres.

Their next check came when they were approaching Pisa, in Italy. A loud report came from the engine. Petrol streamed from the carburettor. There was a flash, and flame crackled along the fuselage. Parer cut off the petrol and threw the plane into a side-slip to keep the flames away from the machine. They fell almost vertically on their side till the petrol on the cowling burnt out, then dropped to an aerodrome which appeared almost miraculously below them. The flames had not weakened the machine and they quickly repaired the jammed needle valve. Over Vesuvius they asked for trouble. McIntosh decided he wanted to photograph the inside of the crater. Parer brought the plane down, only to be caught in a vicious down-draught. The plane fell like a plummet 500 feet towards the crater. McIntosh was nearly thrown from the cockpit. He recorded in his diary that he saw Parer rise a couple of feet from his seat. Parer won control in time to scrape over the rim.

Of this experience Parer wrote: "It seemed as if the machine had been hit a tremendous blow from above. She dropped down as suddenly and helplessly as a feather caught in a sudden downward gust. It seemed only an instant when we were practically over the brink. The petrol in the half empty tanks hit the top with such a thud that it made me fear for the structure of the machine. This showed she must have accelerated downwards and not merely dropped. I grabbed at the seat to hold myself in, but my hands were torn away. It seemed we were being irresistibly drawn into an inferno. The feeling of helplessness was nerve-racking. The engine stopped and started again and pulled us over the top out of the smoke."

From this escape Parer and McIntosh struggled across the Mediterranean to Cairo where R.A.F. men said they were crazy to try to cross the Syrian Desert in a D.H.9. The British pilots regaled them with stories of what Arabs did to airmen who fell into their hands and, when they said they were determined to go, gave them five Mills bombs. "They might come in handy," they said. The Mills bombs did come in handy. Parer and McIntosh were forced down twice in the desert. They spent one night shivering in their plane and found themselves surrounded by Arabs next morning. When one Arab tried to steal a revolver, McIntosh got a Mills bomb from the cockpit. He drew the pin and tossed the grenade over a dune. The explosion put the Arabs to flight. Parer and McIntosh were off before they could return.

The airmen flew through a "wall" of sand near Bushire, the stinging, blinding particles clogging their ears and nostrils. At Calcutta they had just six shillings between them. They had to get cash somehow. Local racing men gave them £400 for stunting over the racecourse, and they earned hundreds more by roaring the full length of Calcutta's main street scattering leaflets advertising tea, oil and other commodities. They left Calcutta for Rangoon on March 25 with £1000 to see them to the end of the trip. Over the Irrawaddy their engine threatened to catch fire again. Parer brought the plane down on a sandbank in the river. They would have stayed there had not natives swarmed from a nearby village. The natives swam to the sandbank and helped the airmen drag the plane across a shallow channel to the river bank. They then slashed a runway through thick jungle, long enough for Parer to make a bumpy take-off.

The airmen ran into a different problem at Rangoon where a Chinese millionaire offered them dowries of 30,000 dollars if they would marry his daughters. Side-stepping this temptation, they set out for Singapore. Nearing Moulmein their engine cut out again and Parer headed for the racecourse to make an emergency landing. Hundreds of natives beat him there. They overran the course leaving clear a circle so small they could not make a normal landing. Parer had to decide whether to plough through the crowd regardless of life or stall down on the small circle and crash. He stalled. When he climbed down from the wreck he thought their flight was over. They had wiped off the undercarriage. The propeller was smashed, the radiator holed and crumpled, petrol and oil tanks were damaged and the fuselage gashed.

Lesser men would have given up. Not so Parer and McIntosh. They were told of a spare propeller at Calcutta and sent for it. They took the battered radiator to Rangoon. Told it was finished, they had another rigged from two motor-car radiators. Local carpenters made a new undercarriage.

Two months later they were off again, only to crash into a ditch near Sourabaya, breaking the propeller and damaging the undercarriage. August 2, 1920, found them on Timor ready for the last 500-mile hop over the sea to Australia. They overhauled their engine and built a raft of a petrol tank and the motor-car tube lifebelts they carried. The first time they tried to take off, the engine cut out. They investigated and found a large piece of perished rubber blocking a petrol pipe. Had the blockage occurred over the sea it would have meant death to them. They took off successfully the second time. The engine that had given so much trouble, now ran smoothly. Their only anxiety was whether they had enough petrol to reach Darwin. They had almost given up hope when they spotted Bathurst Island. A few minutes later they landed in Darwin and were taxiing towards the crowd when the engine stopped. Their tanks were bone dry. They had made it with only a minute to spare.

Sydney gave a great welcome to Parer and McIntosh. Ross and Keith Smith were waiting with the Mayor of Mascot to meet them. They were carried shoulder high to the official tent where they found that enterprising fans had souvenired their buttons, badges, McIntosh's shoe laces and the tail of Parer's tunic on the way. The Governor entertained them at Government House. The battered plane, however, was not to fly into Melbourne. In a landing at Culcairn, her wheels sank six inches into the ground and she turned completely over. She finished the journey on a railway truck and was reassembled on Flemington racecourse for the official welcome by Prime Minister Hughes. The airmen gave Hughes the bottle of whisky from Peter Dawson and themselves received the £1000 consolation prize from an admiring country.

Raymond Parer went on to win Victoria's first aerial Derby, and to pioneer a New Guinea air route, flying supplies and passengers into the Bulolo goldfield.

McIntosh went in for passenger and demonstration flying in Western Australia. He was taking two passengers for a short flight at Pithara on March 28, 1921, when his plane went into a spin and with its engine roaring, crashed into the ground. McIntosh and one of the passengers were killed instantly. Sensation flared at the inquest when witnesses claimed the two passengers were "under the influence of drink" and that two beer bottles were taken from them before they left the ground. Another witness said she followed the flight through field glasses and saw one of the passengers standing in the plane.

McIntosh's manager told the coroner that in his opinion the pilot was interfered with in the air, while another witness referred to stories around the town affirming and denying that McIntosh had been hit by a bottle, and that a man had stood up and overbalanced the machine. The surviving passenger strongly denied these allegations. The verdict was that the "deaths were due to an accidental crashing of an aeroplane, no blame being attachable to anyone." Meanwhile other Australians were planning great flights. The age of the record breakers was near.

Cobham surveys the Empire route

After the thrills of the £10,000 race from England, won by Ross and Keith Smith, Australians became increasingly air conscious. While the Smiths were still winging their way east, young Flight-Lieut. A. L. Long, who had done some test flying and had bombed the enemy during the war, made the first flight across the Bass Strait from Stanley, Tasmania, to Port Melbourne, covering the 180 miles in four hours. Long, who lived at Tea Tree, near Hobart, had to fix extra tanks to his Boulton and Paul biplane, which was more suited to short flights and stunting, to make the distance. He carried about 50 letters to the Governor-General (Sir Ronald Munro-Ferguson), the Prime Minister (Mr. W. M. Hughes), Lord Mayor of Melbourne Aikman, and other notables.

Also in 1919, H. N. Wrigley and Sergt. Murphy made a transcontinental flight from Point Cook, Victoria, to Darwin, in an R.F.C. experimental machine, covering the 2500 miles in 46 flying hours. Other notable flights followed. In 1920, C. J. de Garis, the Mildura dried fruits king, and his pilot, F. S. Briggs, set records with flights across the continent — Melbourne, Perth, Albany, Mildura, Sydney, Melbourne — completing the 6000-mile trip in 5½ flying days. War photographer Captain Frank Hurley took two Curtis seaplanes with him when he led a scientific expedition into the wilds of Papua in 1922. Hurley flew over unexplored stretches of the Fly River, the haunt of cannibal head-hunters.

Distances increased in 1924 when S. J. (later Air Vice-Marshal) Goble and his observer I. E. McIntyre flew round Australia in a short-range Fairey seaplane, covering the 8568 miles in 90 flying hours spread over 44 days. At times, they had to stand watch at night to stop their plane from drifting on the rocks. Later the same year Captain E. J. Jones, with R. H. (Jock) Buchanan as flight engineer and Col.

(Above) The first plane was built in Australia in 1910 by John Duigan, who is shown here at the controls of the historic aircraft.

(Below) William Ewart Hart, the Parramatta (Sydney) dentist, who made the first cross-country flight in Australia and defeated the American "Wizard" Stone in Australia's first air race in 1912.

J.J. Hammond, the first New Zealander to learn to fly, piloting a Bristol biplane at Melbourne in 1911. Hammond tried to interest the Australian

H. C. Brinsmead, Controller of Civil Aviation, as passenger, cut the time for the round-Australia trip to 22 days. In 1925, the Italian Marchese di Pinedo dropped in at Point Cook (Melbourne) in his Savoia flying boat 35 days after leaving Italy on a flight to Japan.

This succession of notable flying feats convinced many Australians that the time for a regular air link with Britain was near. A thrill of expectancy, therefore, ran through England and Australia on June 30, 1926, when Captain Alan Cobham took off from the Medway, near London, in a De Havilland seaplane to survey an aerial route to Australia. Cobham was the ideal man for the job. He had already surveyed routes between England and India and England and Capetown. He was courageous and determined but never reckless. He was not out to set or to break records but to find the best route. He deliberately chose the worst time of the year to find whether regular air services could be maintained in the monsoon period. Cobham carried as mascot a tiny boomerang, inscribed "I go to return", made of British oak by Australian student Roy Bayley. He carried a sea anchor and the latest lifesaving suit lined with a new rubber composition a fifth the weight of cork.

Cobham and his mechanic, A. B. Elliott, made good progress till July 9 when they ran into a dust storm approaching Basra on the Persian Gulf. The dust forced them to descend from 5000 feet to 60 feet. They were finding their way precariously over the swamps when a loud explosion shook the cabin. Cobham thought the rockets had exploded or that Elliott had accidentally fired a revolver. Cobham shut off the engine. "Is there fire," he shouted. "No," replied Elliott, "I am hit." Cobham started the engine again and passed pencil and paper to Elliott. Elliott sent back a note, feebly written. He reported that a petrol pipe had burst and wounded him in the arm and the body. He was losing "pots of blood". Cobham passed him a handkerchief and raced on to Basra. Elliott sank from sight in his cockpit. At Basra he was barely conscious and covered with blood. Cobham, inspecting the machine, found a bullet hole. A swamp Arab had taken a shot at the plane as it passed low overhead. The bullet had ripped through the petrol pipe, shattered the bones of Elliott's arm, passed through both lobes of the left lung and finally buried itself in his back. Elliott died next evening. The death was a great shock to Cobham. He and Elliott were such close friends that, back at their home drome of Stag Lane, they were known as the "Heavenly Twins".

At first Cobham thought of abandoning the flight. The RAF, however, lent him a new mechanic, Sergeant Ward, and they flew on. They survived violent tropical storms which force-landed them, and arrived at Darwin on August 5, 1926. Wheels were then fitted to the De Havilland. Sydney gave Cobham a riotous reception. He was "engulfed in a surging mass of humanity, who cheered and threw hats and everything movable into the air", including the barriers police put there to stop them. Six policemen had to rescue Cobham.

Melbourne's reception was even wilder. Thousands of the 15,000 who assembled to welcome the airmen swept away the barriers and rushed the landing ground as he touched down. Many fled frantically to avoid the whirling propeller. Civil Aviation Controller Brinsmead said he felt really sick. He was sure someone would be killed or injured. The crowd then rushed the official enclosure. They swept away and scattered the official party and trampled the chairs and a temporary broadcasting station underfoot. Someone stole the microphone. Women and children screamed. Police struck out wildly with their fists. Dozens, crushed in the crowd, were treated at a first-aid station while one policeman finished up in hospital.

Alan Cobham and Ward duly flew back to London where "to the roar of the crowd and the shrieking of sirens" they landed on the Thames opposite the Houses of Parliament. Cobham, first to fly there and back, proved that a regular air service between England and Australia was now practicable.

Cobham's successful flights gave an enormous boost to aviation. As far as the all-Red British route was concerned, one major challenge remained. No one had yet flown the 1200-mile Tasman between Australia and New Zealand. Naturally, New Zealanders wanted to be first to fly to their native land. The whole country applauded, therefore, when three New Zealanders, John Robert Moncrieff and George

Hood, Air Corps reserve officers then working as motor mechanics, and Territorial Air Officer Capt. Knight, decided to win the honour for New Zealand. Moncrieff, the pilot, would have liked a sea-plane but could not raise the money. Instead they bought an American high-wing monoplane from the firm that had built Lindbergh's trans-Atlantic plane, *Spirit of St. Louis.*

They assembled the plane at Point Cook near Melbourne and were all set to go when Australia's Prime Minister S. M. (now Lord) Bruce banned the flight. He pointed out that Commonwealth regulations would not permit land planes to fly more than 50 miles over the sea. There is reason to believe that he did not think the plane adequate or the crew experienced enough to make the flight. New Zealand Premier Coates then intervened and pointed out that his experts had passed the plane as capable of the flight. Prime Minister Bruce withdrew his protest. The airmen decided that, to conserve petrol, only two should fly. Hood and Knight tossed, at Wentworth Hotel, in Sydney. Knight called heads. Tails came up. Hood, a war veteran with a wooden leg, did a dance of victory.

Calling their plane *Aotea-Roa* (Land of the Long White Cloud), the Maori name for New Zealand, Moncrieff and Hood left Richmond, near Sydney, on January 10, 1928. Neither could operate a wireless set. All they had was an automatic transmitter which sent "whines" every 15 minutes to let listeners know the plane was still in the air. If they came down, they could not send an SOS or notify their position. All New Zealand was certain the gallant men would win. The "whines" were heard regularly. Two ships reported the plane had passed overhead. Thousands assembled at the Trentham racecourse, Wellington, to greet them.

The signals came through steadily till the plane was estimated to be within 200 miles of New Zealand. Then they ceased. Anxiety spread through the waiting crowd. Rumours were rife. The plane had crossed the coast; it was nearing Wellington; it had landed safely elsewhere. The crowd lingered till midnight when Mrs. Moncrieff, standing with Mrs. Hood and other relatives, looked at her watch and said quietly, "Their petrol is out."

An intense search was immediately instituted. The warships *Diomede* and *Dunedin* and a tug searched the area whence it was thought the last signal came. A storm blew up and battered the ships. Reports, some of them hoaxes, declared that the plane had been heard over widely scattered parts of New Zealand. Some papers published that the men were safe. Capt. Knight thought the fliers might have landed in the Rimutka Ranges which Hood knew well. Scores of volunteers, led by mountain climbers, struggled through thick mountain bush in search of a crashed plane without avail. The fate of Moncrieff and Hood remains a mystery. Most believe they crashed into the sea within 100 miles of New Zealand. Some believe their plane may still be rotting on a lonely mountain side.

Perhaps it was this tragedy that made New Zealand so severe when 24-year-old New Zealander Raymond G. (Ron) Whitehead and Ballarat-born Rex Nicholl set off on November 22, 1934, to fly the Tasman in an old Puss Moth which had already been condemned by the Civil Aviation authorities. By then the Tasman had been crossed many times, first in 1928 both ways by Kingsford Smith, Ulm, Litchfield and McWilliams. In 1934, Whitehead and Nicholl decided they would cut the record, then ten hours, in their Puss Moth jalopy. They removed two of the three seats to fit extra petrol tanks. This obliged one to squat between the legs of the other when they took off secretly, in defiance of the official ban, on November 22, 1934. The Puss Moth's compass was faulty. For oil pipe it had a length of hose into which oil was poured from time to time through a funnel.

Despite these handicaps Whitehead and Nicoll made New Zealand though slightly off course. They came down on a lonely beach and flew on to Mangere aerodrome, Auckland, next day, just as planes were setting out to search for them. The national rejoicing was tempered somewhat when the two airmen were summoned for flying an unregistered aircraft, flying in an aircraft not possessing a certificate of airworthiness and failing to carry documents prescribed in the regulations. They were liable to six months' imprisonment, a fine of £200, or both. The magistrate was lenient. He found the first charge proved, but, as this was the first case under the new regulations, freed the men and dismissed the other charges. The Governor-General, Lord Bledisloe, scotched the subsequent controversy by declaring bluntly that, far from earning them a rebuke, their feat should "stimulate the imagination and zeal of other young New Zealanders to emulate them".

Back in 1928, the tragedy of Moncrieff and Hood had put a damper on civil flying. Aviation badly needed a lift. It was given this fillip almost at once by young Australians Bert Hinkler, Charles Kingsford Smith, and C. T. P. Ulm, who were about to annihilate distances.

――Chapter 8――

Hinkler flies alone

One woman and three men watched as a tiny Avro Avian biplane took off from Croydon Aerodrome, near London, at dawn on February 7, 1928, and headed for Australia. The woman was Mrs. Nance Hinkler, whose husband, Bert, was flying the midget plane that vanished so swiftly in the mist. Herbert John Louis Hinkler, D.S.M., was determined to be the first to fly solo to Australia. He intended also to slash the record for the 11,500 mile journey still held by Ross and Keith Smith and to set up others along the line.

Few, apart from Nance Hinkler, believed he could do it. Most considered him crazy to attempt the flight in such a flimsy plane. The Avro Avian was one of the babies of aviation. Its fabricked wings had a span of only 30 feet. They were made to fold back so that one man could wheel it into a garage. Total weight was 900 lbs. The air-cooled engine had only four cylinders. Speed was 105 miles an hour. Experts doubted if such a frail craft could survive the sandstorms and tropical tempests that plagued the three crews, who at much peril had already flown the route. They did not know Bert Hinkler.

Bert Hinkler had always been wild to fly. As a youth in Bundaberg, Queensland, he made man-carrying gliders. He became mechanic to "Wizard" Stone, the American pioneer airman, who barnstormed round Australia and New Zealand before World War I. In the war, Hinkler was an air gunner, then a pilot in France and Italy. He entered for the £10,000 race from England to Australia in 1919 but failed to get a plane. Hinkler was determined, however, that he would fly to Australia. As a preliminary he made record flights to Turin and was a partner in an ill-fated bid to fly non-stop to India. During a trip to Australia he caused a sensation by flying non-stop from Sydney to Bundaberg in record time and landing in the street outside his mother's house.

Bert Hinkler, therefore, was both an experienced pilot and expert mechanic when, at last, he decided to "go it alone" on a flight to Australia. Nothing went smoothly for him. He tried to get backing from newspapers but "the response was about as good as if I had been trying to sell rotten fruit". Hinkler had only 2½ hours' sleep before setting out on the record flight. His wife's car broke down as she was motoring up from their home near Southampton. She did not arrive till 1.30 a.m. and then, according to newspaper reports, they could find accommodation only in the aerodrome sick bay.

Hinkler was up at five. The weather was unfavourable for flying. He pondered the problem a few moments and then said, "I'll chance it." Mrs. Hinkler, a former hospital sister, an Avro manager and two others saw him take off. Nance Hinkler did not betray the slightest anxiety. She was only sorry she was not going too. Thirteen hours later Hinkler touched down in moonlight at Rome. He was promptly arrested and marched to the lock-up between two soldiers for landing on a military instead of a civil airfield. In his anxiety he slept only four hours, so again was not at the top of his form when a Consul rescued him next morning and sped him on his way.

Hinkler landed twice in the desert in the Middle East, inflated his rubber dinghy and slept in it. Arabs who came up during the night were friendly and helpful. One party cleared a path through camel thorn so he could take off. By now Hinkler was suffering from the excessive heat and had bouts of cramp. To stave off depression, he sang and talked to himself. The world, however, was beginning to take notice. It was plain that if he kept it up Hinkler and his midget plane would slash Ross and Keith Smith's time almost in half.

A Sydney newspaper gave a greater sporting aspect to the flight by announcing that Hinkler had taken out a policy with Lloyds under which he might "net a considerable sum". "For a premium of £150," said the report, "Lloyds has undertaken to pay a reward increasing in geometrical progression for each day taken off Sir Ross Smith's record of 28 days. If Mr. Hinkler reaches Australia in 27 days, he receives £1, in 26 days £2, in 25 days £4. In the event of his occupying 20 days he receives £128, 19 days £256, 18 days £512, 17 days £1024. Hinkler is aiming at the last named sum."

Over Arabia Hinkler was five days ahead of Ross Smith's time. On the way to Karachi, he had a few anxious hours racing a leak in a petrol tank. He reached India on the eighth day, establishing a new record. The world's newspapers were lyrical over the feat. One called him "Hustling Hinkler". Some, however, still croaked. "He can't keep it up," they said. Hinkler flew on undaunted. On the way to Rangoon he reported many fires in the jungle, "which cover the ground like the matted hair of the Fijian". By now, as the midget plane cut back the miles, elated fans in Australia were working out how much he stood to win in his gamble with Lloyds. Hinkler made an emergency landing in a jungle clearing and switched 50 miles out of course to dodge tropical storms on the way to Singapore. There his midget plane bogged. He had to get onlookers to give him a shove before he got off next day.

Though the Commonwealth Government affected to hold aloof from such wild enterprises, a warship happened to be cruising in the Timor Sea when Hinkler took off on February 22 for the final 850-mile hop over water to Australia. Some were beginning to abandon hope when a man with a telescope saw what looked like a silver-backed beetle gliding through the clouds at tremendous height toward Darwin.

A few minutes later Hinkler landed to a hero's welcome, 16 days after leaving London. The flight had cost him £55. From then it was one long triumph with only one hitch. On the way to Bundaberg, Hinkler was forced down and had to spend a night in the open in his inflated dinghy. Next day, at Longreach, an over-ardent fan tried to souvenir his goggles, while admiring mothers placed their babies on the wings of his plane, which did not do the wings much good.

Bundaberg and Brisbane went wild over Hinkler. In Brisbane, his midget plane, wings folded sedately back, was mounted on a lorry to share the honours in a triumphal procession. Between 80,000 and 100,000 people welcomed him at Sydney, where Sir Keith Smith gave him the proud title of "Australia's Lone Eagle". Every capital city feted Hinkler with luncheons, dinners, receptions and gifts. The whole world went Hinkler mad. Everything that moved quickly was said to "hinkle". Everyone sang the song "Hustling Hinkler". Mothers named their babies Herbert. Flappers wore Hinkler hats patterned on a flying helmet. A movie company offered Hinkler £100 a week for 12 weeks to star in a film, while another firm promised £1000 if he would undertake a lecture tour.

Punch capped it all with the famous cartoon of a kangaroo dancing wildly on the shore waving the Australian flag at a tiny biplane, with the caption "Hinkle, Hinkle, little star! Sixteen days — and here you are!" The only sour note came from Perth, where certain religious leaders reproached Hinkler for landing there on the Sabbath and turning the sacred day into a carnival. Roman Catholic Archbishop Clune gave a touch of humour to the controversy, however, by remarking drily that "While Hinkler had made a remarkable flight, he had also performed a sort of miracle in Perth. For the first time in its history, they had seen the spectacle of all the inhabitants gazing earnestly to heaven for an hour or two".

Meanwhile his sporting fans worked out that by completing the flight in 16 days, Hinkler had won £2043 in his gamble with Lloyds. A groan of sorrow rose, however, when the underwriters reported blandly that, although Hinkler began negotiations for such a policy, he had not completed them. Hinkler confirmed this. He had kept the premium money for expenses of the flight.

His admirers, however, made sure Hinkler did not lose by it. Prime Minister S. M. Bruce's Federal Government forgot its edict against wild enterprises and voted him £2000. The Queensland Government gave him £500. Clubs and other organisations all over Australia collected to show their admiration of one of the world's pluckiest airmen. When the furore ended it was estimated that an admiring country had subscribed over £10,000 for the modest hero who had flown to Australia on a shoestring.

Hinkler went on to greater triumphs. In November 1931, telling no one of his plans, he flew from Toronto (Canada) in a small De Havilland Puss Moth, pushed on via New York, Cuba, Jamaica to Fortaleza, in Brazil, where he was arrested for landing without a viza. On release, still keeping his plans secret, he flew to Natal, in Brazil. On November 23 he took off again and was seen heading east across the South Atlantic. For 24 hours nothing was heard of him. Then the news flashed across the world that Australia's Lone Eagle had landed at Bathurst, on the West African Coast. Bert Hinkler had made the first west-east crossing of the South Atlantic, the first solo flight across the South Atlantic and the first transatlantic flight in a light aeroplane.

This won him great praise but little wealth. With the slump setting in, Hinkler needed money for the production of a small amphibian plane he had designed and called the Ibis. Hoping the feat would win him financial backing, Hinkler set off in his Puss Moth on January 7, 1933, to beat C. W. A. Scott's time of eight days 10 hours to Australia. Thick fog shrouded the aerodrome. The cold was intense. Snow was falling in the Alps and Italian Apennines that straggled across his route. Into them Hinkler vanished. Nothing was heard of him till April, when a band of Italian charcoal burners stumbled across the wreckage of an aeroplane 3000 feet up in the mountains. Near it was Hinkler's body. Australia's Lone Eagle had died as he had flown — high and alone.

— *Chapter 9*

Across the Pacific

Three young Australians stepped briskly ashore in San Francisco on August 5, 1927, and announced confidently that they intended to make the 7000-odd-mile flight across the Pacific to Australia, a feat no one thought possible. The Australians were Charles Kingsford Smith and Keith V. Anderson, both fighter pilots with kills in France, and Charles T. P. Ulm, who served in Gallipoli and Flanders and was now determined to carve for himself a career in aviation.

For seven years, ever since Prime Minister W. M. Hughes banned him from the first great race to Australia, Kingsford Smith had dreamt of being the first to fly the Pacific. The dream grew as he walked the wings and hung by the knees from the axles of flying planes to provide thrills for Hollywood films. It built into an obsession as he barnstormed Australia and flew passengers on one of the continent's first airlines.

Kingsford Smith won a sympathetic hearing from his fellow pilot Keith Anderson with whom, for a spell, he went into motor transport, servicing the outback. Their hearts, however, were in flying. His new chance came when Charles T. P. Ulm, a sound organiser, asked them to help him found an airline. The plans fell through; all that was left was Kingsford Smith's great dream of a trans-Pacific flight. Unfortunately, they had little money. They needed very substantial backing and no one was willing to advance hard cash to three young men for such a crazy undertaking.

They decided, therefore, to stunt their way into public confidence. To do this Smithy and Ulm flew 7500 miles round Australia in a Bristol tourer, slashing the record from 22 days to 10½. Premier Lang was there to meet them when they flew in triumph back to Sydney. The actress Nellie Stewart gave them a kiss apiece. Every Australian praised them. If they could fly 7500 miles round Australia, maybe they could fly the slightly shorter distance across the Pacific, despite the long hops over sea. Premier Lang offered to back them up to £3500.

Thus encouraged, Smithy, Ulm and Anderson sailed for San Francisco, well knowing they'd need a lot more money. San Francisco then was in an aerial frenzy. Charles Lindbergh had flown the Atlantic solo. Hardly had he landed when James Dole, the Hawaii pineapple king, offered 35,000 dollars in prizes for a race from California to Hawaii. Crazy young aviators rushed to enter. Some could hardly fly. Few could navigate. Their planes were flying crates.

Hoot Gibson, the film star, entered a crazy triplane of wood, canvas and wire which crashed in San Francisco Bay on its first test flight. Lindbergh was horrified; he called the race sheer suicide. Only the most skilled navigators could hope to hit Hawaii, a pin-point in a vast ocean, he said. A hopeful owner asked Charles Kingsford Smith to pilot one of the planes. "I doubt if the damned thing will ever get off the ground," said Smithy and went back to his trans-Pacific plans.

Several airmen were killed before they reached the starting post in the Hawaii Pineapple Derby. Seven vanished in the race, including 22-year-old Miss Mildred Doran, "the prettiest little pigeon on wings". Only two of the eight starters limped into Wheeler Field, Hawaii. Another pilot lost his life searching for survivors.

The mass tragedy knocked all the aerial ballyhoo out of San Francisco and wrecked Smithy's hopes of getting backing for his trans-Pacific flight. If Americans couldn't even reach Hawaii, what chance had he of flying all the way to Australia. He was mad to think of it. It couldn't be done. No one in his senses would advance money on such a crazy venture. Eight months of frustration followed for Smithy, Ulm and Anderson.

Their first break came when another Australian, Adelaide-born G. H. (later Sir Hubert) Wilkins offered to sell them a sturdy plane said to have been built from three-engined and single-engined Fokkers he smashed up in the Arctic. He offered them the plane without engines or instruments for £3000. Wilkins had won world fame by flying

550 miles into the Arctic, landing on ice, taking ocean soundings and trudging 18 days over ice to safety after being forced down in the white wilderness. He needed money to buy a new plane in which to fly over the North Pole from Alaska to Spitzbergen in Europe.

By now Smith, Ulm and Anderson could not afford £3000. Wilkins fixed that by accepting a down payment of £1500. A wealthy Australian business man, Sidney Myer, passing through San Francisco, gave them £1500 towards the three Wright Whirlwind engines which they speedily installed. A period of sheer heartbreak followed. As the lessons of the Pineapple Derby sank in, messages came from all over the world pleading with them to abandon the flight. Even the Government had second thoughts. Money tightened. Ulm recorded that at one time they were down to their last 16 cents. They did not know where their next meal would come from. Anderson packed up and went back to Australia.

In a desperate bid for prize money and prestige, Smith attacked the world endurance record of 52 hours, 22 minutes in the air, then held by Germans. The *Southern Cross,* as he now called the plane, was dangerously overloaded with petrol when he and another pilot took her up, only just bumping clear of obstacles at the end of the runway. For two weary days and nights they circled San Francisco Bay till, at the second dawn, their fuel ran out. They failed by two hours to break the record.

Numb and exhausted, Smithy staggered disconsolately from the plane. There would be no prize-money, no public praise to encourage moneyed men to back them. The two Australian airmen plumbed the depths of despair. Friends still urged them to give up, which, with all the bills they owed, would mean bankruptcy. They tried to sell the *Southern Cross* to an oil company but failed.

Creditors were on the point of seizing the machine, when out of the blue, a banking friend introduced them to the wealthy American industrialist, Captain G. Allan Hancock, himself an expert navigator. Hancock took Smith and Ulm for a cruise on his luxury yacht. He quickly realised they were not crazy fools. They knew what they were doing and had taken every precaution against disaster. He asked Smith

how much he wanted. "£3200," came the reply. "I'll buy
your plane for £3200," said Hancock, "and you shall fly it to
Australia."

Thus it came about that on May 31, 1928, Kingsford
Smith and Ulm set out on the most hazardous flight to that
date, the crossing in three great hops of the 7000-odd miles
of the Pacific Ocean. With them were American Navy Captain
Harry Lyon, a son of a rear-admiral (navigator), and his
friend Jim Warner (wireless operator). On the first leg, 2400
miles to Honolulu, they were guided by radio beacon. Their
only mishap came when their radio transmitter failed, putting
them out of touch for a while with an anxious world. They
landed at Wheeler Field near Honolulu, 27 hours 27 minutes
after leaving Oakland.

The second leg entailed the longest flight yet attempted —
3138 miles over sea to Suva. Croakers doubted openly if they
could do it. Some even forecast they'd be down in the sea
long before they reached Fiji. Ships along the route,
including two New Zealand warships, were ordered to keep a
lookout for them. This time the *Southern Cross,* dangerously
overloaded with petrol, ran into tempest. Smithy and Ulm,
taking turns at the controls, had to weave, soar and swoop to
dodge the storm clouds. They drove straight through squalls;
lightning ripped around the cabin; air currents tossed the
plane about. The whole western world waited tensely. There
was general relief when, after 36½ hours, the *Southern Cross*
touched down on the pint-sized oval at Suva. Smithy and
Ulm had made the longest ocean flight on record. A thrill ran
through Australia. Prime Minister S. M. Bruce stopped
parliamentary business to announce, "The airmen have
landed in Fiji".

The final leg, 1500 miles to Brisbane, should have been the
easiest, but turned out to be the worst. The airmen ran into a
tempest they could not dodge. Some of the bumps were so
violent that it needed the strength of both men to keep the
plane under control. Cotton wool was useless to keep out the
noise, so they plugged their ears with plasticine. For miles all
they could see was the flash of lightning and a rushing
cascade of water. Flying blind they crossed the Australian

coast about 110 miles south of Brisbane, where they landed after a nightmare flight of more than 21 hours.

Brisbane gave them a tumultuous welcome. Waiting for them, also, was a cable from America giving them the plane and announcing that all their debts totalling nearly £10,000, had been cancelled. More than 200,000 greeted the four at Mascot (Sydney). One lyrical writer claimed that the "block of noise" from ships' sirens, motor horns and cheering people would make "Gabriel's trumpet sound like a tin whistle competing with a barrage". He described the Governor-General (Lord Stonehaven), Prime Minister Bruce and others in the official party as standing "crushed like fossils in pre-Cambrian limestone" on their dais, and deplored the "trampling of toppers" and "mangling of monocles" in the crush. Finally, he wrote gloomily that there'd be 100,000 thick heads and sore throats next day.

Everyone clubbed to honour Smith and his companions. The Government gave them £5000. Admiring fans subscribed to bring the total to £20,000. Three months later, Kingsford Smith and Ulm were in the air again, out to notch another "first" — the first flight from Australia across the stormy Tasman to New Zealand and back. This time they had H. A. Litchfield as navigator and T. H. McWilliams, superintendent of the Union Steamship Company's wireless school in New Zealand, as wireless operator. The flight was considered so hazardous that the Commonwealth Government arranged for the warship *Anzac* to be cruising along the route while they were in the air.

Smithy and Ulm planned to leave Richmond, N.S.W., at 6 p.m. on Saturday, September 2, 1928, fly through the night and land at Wigram aerodrome, near Christchurch, early the next morning. The proposal to land on Sunday brought a storm of protest from Christchurch clergymen. A deputation of indignant churchmen waited on the Rev. J. K. Arthur, Labour Mayor of Christchurch, protesting against "the sanctity of the Sabbath being set at naught". It was an insult, they said, to the whole Christian community and "as the flight had been arranged to advertise a new brand of petrol it seemed like a gross commercialisation of the Lord's Day."

Mayor Arthur agreed warmly with them. He stated

publicly that he refused to take part in a welcome to the flyers on a Sunday. He sent a stern cable to Squadron-Leader Kingsford Smith. "Christchurch churchmen strongly protest against plans involving arrival on Sunday. I support the protest. Cannot departure be delayed." Departure was delayed — by gales which blew up in the Tasman. The *Anzac*, which had already put to sea, was recalled to port.

The gales raged till September 10 when reports came that the weather was practically clear and as favourable as could be expected at that time of year. After farewelling a host of friends, Kingsford Smith, Ulm and their crew, took off in the *Southern Cross* and sped east. The warship *Anzac* which had left port six hours earlier cruised along their route.

Whoever reported weather over the Tasman as clear was slightly off beam. For two hours they battled through storm. Lightning played round the plane and put the wireless out of order for an hour. The weather for the rest of the flight was boisterous. They had many bumps, the biggest dropping them 300 feet.

Fourteen hours after leaving Richmond, the *Southern Cross* landed at Wigram where 30,000 excited fans waited. The authorities thought there might be trouble so they'd called out the Territorials to help the police keep order. They didn't have a chance. The crowd knocked down the barriers, brushed aside the troops and mobbed the airmen. Sir Heaton Rhodes was able to welcome them on behalf of the Government, but Mayor Arthur, who had relented and come to greet them as chief citizen of Christchurch, was cut off by the crowd and could not be found. Some of the crowd scratched paint from the plane for souvenirs. Smithy lost his helmet — to a too-eager souvenir-hunter, reported one paper sternly. Navigator Litchfield was cut off from his mates in the crowd and had to get a lift in a furniture van to the reception in Christchurch.

A huge crowd shouted for Kingsford Smith and Ulm at the hotel where they stayed. Cheers rose when they appeared on the balcony. Smith cupped his hands round his mouth. "We're sorry we can't ask you all in for a drink," he shouted. "We would have liked to reach Christchurch on Sunday, but couldn't make it." A great roar greeted this remark. "Did you

(Above) Bleriot monoplane of Maurice Guillaux taking off from Wangaratta, Victoria, on the first mail flight.

(Below) Maurice Guillaux who flew the plane which carried Australia's first Air Mail, from Melbourne to Sydney in July, 1914.

(Right) The Vickers Vimy which won the England-Australia air race in 1919. In the cockpit are Ross Smith, Keith Smith and W.M. Shiers.

get the mayor's cable?" bellowed one man in the crowd. "Oh, yes, we got it," replied Smith with a grin.

Next day an appeal went out for Smithy's helmet. "Smith does not want the helmet particularly," reported one newspaper, "but he wants very badly the figure of a black cat which was clasped to it. To him this is a very valuable mascot and he is offering a reward of £10 for its return." Two days later Smithy got his helmet and mascot back. A young woman who had found the helmet and kept it as a souvenir returned it, asking no reward. Smith promised to use the helmet on the flight back to Sydney, then autograph it and post it back to the finder in Christchurch. The New Zealand Government made a grant of £2000 for the flight.

The return flight from Blenheim, New Zealand, to Richmond was a nightmare. Kingsford Smith, Ulm and their crew set off soon after 3 a.m. on October 14, 1928, and headed west, buffeted by gales and harassed by fog. This time two warships, *Australia* and *Anzac,* patrolled part of the route. Despite the weather, crowds flocked to Richmond to greet them. The Richmond-Windsor road and the approaches to the aerodrome were choked with vehicles. At dusk, the crowd had built up to 25,000. Everyone settled down happily in the darkness to await the return of the airmen.

As the hours dragged slowly by, some anxiety was felt. Every preparation was made, however, for a night landing. More than £1000 had been spent fitting up some of the most up-to-date floodlights, searchlights and flares which, it was confidently expected, would turn the whole area of 2000 square yards into "a fairyland of light". One huge sun arc light, described as the most powerful in Australia, could be seen 70 miles away. There was much consternation, therefore, when, with the night at its blackest, the electricity supply failed and plunged the whole area and the crowd into complete darkness. Official chagrin turned to dismay, then to despair. The *Southern Cross* was due at any moment. While the electricians toiled, officials debated the hazards attending a landing in the dark and danger to the crowd if a forced landing, through lack of fuel, became necessary.

A rumour then flashed through the crowd that the *Southern Cross* had been sighted at Barrenjoey, a few miles away, and was due to land at any moment. Officials and police acted promptly. They tore down a section of the fence skirting the enclosure and appealed to drivers of motor cars to drive onto the aerodrome and illuminate a runway with their headlamps. Drivers drove their cars at furious pace across the aerodrome and parked with their headlights facing south. Five hundred were in position before it was realised that the alarm was a hoax.

Meanwhile, the airmen had reached the coast at Nobby's Head, Newcastle, only to be lost in fog in which they cruised blind for two hours.

By the time they reached Richmond, electricians had got the arc lamp back into commission. The plane came overhead "like a great bat in the darkness" and touched down at 2.15 a.m. helped by the arc light, motor-car lamps and a dozen petrol flares. They had only three gallons of petrol left, enough to keep them in the air for no more than ten minutes.

The airmen left the plane to cries of, "Good on you, Smith" and "For they are jolly good fellows". Owing to the two hours lost in the fog, they had covered 1650 land miles and had taken 23 hours for the flight. Smithy said a more suitable plane would have to be evolved before they could make the crossing regularly.

Two years were to pass before young Guy L. Menzies, son of a Drummoyne (Sydney) doctor, and holder of a B licence, made the first solo crossing of the stormy Tasman. Menzies, aged only 21, and a partner owned the *Southern Cross Junior,* the Avro Avian in which Kingsford Smith had broken Bert Hinkler's England-Australia record. Menzies told no one he proposed to fly to New Zealand. Relatives and friends assembled at Mascot aerodrome to farewell him on what they thought was a flight across the continent to Perth in Western Australia.

After he entered the cockpit, Menzies beckoned his brother Ian to the plane and handed him a number of letters addressed to his parents and to friends and endorsed "Not to be opened till after the take off". Watching friends were surprised when, instead of turning west, he headed east over the sea. They then opened their letters and found he was on the way to New Zealand. Amazement gave way to anxiety.

Some considered him a trifle foolhardy, but there was nothing they could do about it, except notify New Zealand that he was on the way. Great excitement flared across the Tasman. It gave place to anxiety when young Menzies was considered overdue.

Menzies got safely through, however. When he reached the New Zealand coast, he cruised north along the beach looking for a suitable landing place. With petrol running low, he turned inland and seeing what appeared to be a smooth patch in the bed of the Wanganui River landed with only half a gallon of spirit left in his tank. The spot he chose was marshy. The wheels sank in the ground and the plane turned over. The crash was heard half a mile away. Farmers rushing to the scene were relieved to see Menzies crawl out unhurt. He had taken 12¼ hours for the 1190-miles flight. Menzies explained that he had told no one of his plans and had left the letters so others could not be blamed for encouraging him in his adventurous flight.

Meanwhile another Australian had flown into history. A few weeks before Smith and Ulm crossed the Pacific, G. H. (later Sir Hubert) Wilkins had achieved his great ambition by being the first to fly over the Arctic ice from Alaska to Norway, thus pioneering the polar route.

— *Chapter 10* ————————————————

Over the top of the World

George Hubert Wilkins of Adelaide, notched another "first" for Australia when, in 1928, with American Ben Eielson as pilot, he flew over the top of the world from Point Barrow in Alaska to Spitzbergen in the Norwegian isles north of Finland. Born in 1888 at Mount Bryan East, South Australia, and educated in engineering at the Adelaide School of Mines, Wilkins led a life of constant adventure. As far back as 1916, when he ended three years as photographer and second-in-command of Vilhjalmur Stefansson's Canadian Arctic Expedition to join the Australian Flying Corps, he had dreamt of exploring the icy Arctic wilderness in aeroplanes, chancy and unreliable though they then were. With Stefansson, Wilkins learned to live like an Eskimo. He tramped more than 5000 miles with the expedition and became as much at home on ice as an Arctic seal.

From the Flying Corps Wilkins was seconded as official photographer to the Military History Department, in which capacity he was wounded several times taking photographs in battle. Having won the M.C. and bar, he flew with Val Rendle, Reg Williams and Gar Potts in the first England-Australia race in 1919, navigating the Blackburn Kangaroo aircraft which crashed in Crete. In 1920, Wilkins was second-in-command of a not-too-successful British Imperial Expedition in which he helped survey part of the Antarctic coastline, then joined Sir Ernest Shackleton's last Antarctic Expedition as naturalist and photographer. Switching back to the sun, he roamed tropical Australia for two adventurous years collecting natural history specimens for the British Museum.

Now (1926) aged 38, he returned again to his earlier dreams of exploring the Arctic from the air and of being the first to fly from America to Europe by the polar route. With his own savings and financial help from Australian friends,

Wilkins sailed for America to seek backing there for his polar plans. The City of Detroit voted him a substantial sum. With this and 25,000 dollars from a North American newspaper *Alliance* for exclusive news coverage of the flight, Wilkins had enough to buy two Fokker aircraft — one tri-motor, the other single-engined. His old Arctic commander, Vilhjalmur Stefansson, put him in contact with Dakota-born 28-year-old Carl Ben Eielson, a crack young "bush pilot", thus forming a partnership that ranks high in Arctic exploration and aviation history.

At the beginning it looked as though their plans were hoodooed. Both aircraft crashed during tests. Before they could be repaired the Arctic fogs set in and forced Wilkins to postpone the big attempt for a year. To make things worse, their press agent, Palmer Hutchinson, walked into a spinning propeller and was killed, and Wilkins himself broke an arm man-handling an aircraft out of a snow drift. All they could hope to do that first year was to ferry stores and fuel to their base at Point Barrow, by dog team, snow tractors and by air, in which they flew over rugged snow-bound mountains, never before crossed by man.

In March 1927, Wilkins and Eielson, with two new planes were ready for another aerial attack on the Arctic. Admiral Richard Byrd, the American, had forestalled them to some extent by flying from Spitzbergen to the North Pole and back in 1926, but the whole of the Arctic north of Alaska was still unexplored, and no one as yet had flown from one Continent to another over the North Pole. Most experts declared it just could not be done.

Wilkins decided to start his 1927 programme with a 600-mile flight towards the pole to see if there were any substantial stretches of land or islands still waiting to be found in the vast wastes of ice and snow. They took off from Point Barrow with a blizzard on their tail and were going well 550 miles out when the engine faltered. Rough ice lay beneath them. There was nothing much they could do. They just had to risk a landing. Eielson eased the plane down. The ice held. She skidded along on her skis and came gently to rest in the first landing ever made on Arctic ice.

While Eielson fixed the engine, Wilkins cut a hole in the ice and made a sounding with a listening device and dynamite. He found the ocean more than three miles deep at that point. They took off and again were forced down. They worked on the engine for some hours in sub-zero temperatures in which four fingers on Eielson's left hand froze. Black clouds blocked the light when at last they were airborne again. Now they were battling a gale-force headwind without a chance of getting back to base before their fuel ran out. They were not surprised therefore when the engine cut out. Eielson felt his way down to a landing. The wing tip struck an obstruction and spun the plane into a snowdrift.

Neither Wilkins nor Eielson was hurt. For five days they sat out the blizzard in the cabin of their plane, resting snugly in their sleeping bags and living comfortably on their emergency rations of pemmican, biscuits, chocolate, nuts, raisins and malted milk tablets. Then, with a pocket compass and two watches, they set out to walk nearly 100 miles across the ice, back to civilisation. Carrying rifles and with 30 lb. of food bundled with their instruments into sleeping bags, they trudged, stumbled and, in some places, crawled southwards over the ice. Eielson, in constant pain from frostbite, plodded doggedly along with Wilkins. Tragedy was near when the ice gave way under Wilkins while he was skirting a lead of open water. He managed to scramble, soaked to the skin, on to firm ice. Only his knowledge of Arctic lore prevented him from freezing to death. After 13 days of privations, the two men reached a trader's house on Beechey Point, a few miles from Port Barrow. Eielson was taken at once to hospital where surgeons amputated a finger from his frost-bitten hand.

Again Wilkins was forced to postpone his main objective, a flight over the Pole from Alaska to Northern Europe. During the winter he went to San Francisco where, as previously recorded, he met Charles Kingsford Smith and C. T. P. Ulm and sold them the sturdy plane he had built from the Fokkers he smashed up in the Arctic. In San Francisco, Wilkins saw and fell in love with a trim little Lockheed Vega monoplane, whose sleek, bullet-like body offered a minimum of wind resistance and was just the plane for Arctic exploration. The money he got from Smithy and Ulm helped him buy a Vega. In April 1928, he and pilot Ben Eielson were poised at Point Barrow for the great adventure.

As usual, nothing went smoothly. They had to marshal scores of Eskimos to clear a runway through the snow. Twice, the Vega, greatly overloaded, refused to rise. They had to wait for a head wind before, on April 15, the Vega rose smoothly from the snowy runway and headed for Spitzbergen, 2500 miles away. Weather was mixed. After seven hundred miles they ran into black clouds and were lucky to be able to check their position by glimpses of Greenland. mountains. Next they encountered a blizzard which cut visibility to 100 yards. Nearing the end of their journey, they narrowly escaped crashing into the sea. Fuel, now, was dangerously low. Wilkins guided Eielson down to a small strip of ice, confident they had succeeded in their mission and were somewhere in the Spitzbergen group of islands. The 2500-mile flight across the top of the world had taken 20 hours, 20 minutes.

For five days, however, they were stormbound in the cabin of the Vega. Then they took off and, a few minutes later, saw the masts of the wireless station at Green Harbour, where the Norwegian operator and his crew found it difficult to believe they had crossed the Arctic from Alaska.

Honours poured on Hubert Wilkins. King George V knighted him. Scientific bodies and ordinary folk all over the world hailed him as a great explorer and aviator. The newspaper tycoon William Randolph Hearst advanced 25,000 dollars towards the cost of an aerial exploration of the Antarctic. With Ben Eielson as first pilot and another famous airman, Joe Crosson, as second, Wilkins arrived with two Vegas at Deception Island and, in November 1928, made the first flight over Antarctic snows. In December he was off again flying 600 miles out over the high plateau of Graham Land where the foot of man had never trod. On a spot left blank on the map he plotted plateaus, bays, channels and chains of islands none knew existed, returning safely through a fierce storm to base at Deception Island. Not for nothing were Wilkins and Eielson known as "The men who always came back".

Romance came to Sir Hubert Wilkins in 1929 when

he married the beautiful Australian actress Suzanne Bennett in America.

Wilkins now took two years off from flying to embark on another adventurous project, the possibility of exploring beneath the Arctic ice cap in a submarine. For the token price of one dollar, the United States Navy let him have an out-of-date submarine, about to be scrapped. Lady Wilkins christened it *Nautilus*. Wilkins fitted the submarine with runners and shock absorbers on top of the hull and with a hollow drill in the conning tower to enable them to cut through the ice for air. In *Nautilus*, Wilkins hoped to sail from Spitzbergen under the North Pole to Alaska in 42 days.

The *Nautilus* project was dogged by misfortune. She made one or two trial dives under the ice to prove it could be done before the experiment was abandoned. Wilkins' faith in submarine voyages under ice was vindicated when American nuclear-powered submarines made such trips after World War II.

For the next six years Sir Hubert Wilkins was manager of the Lincoln Ellesworth Antarctic Expedition. In 1937 he was called back to active flying to search for Russian airman Sigismund Levanevsky, the Soviet Lindbergh, who vanished with five comrades on a flight over the North Pole. Before the search was called off, Wilkins had flown 44,000 miles and searched 170,000 square miles, 150,000 of which had never been surveyed before. Altogether Hubert Wilkins made more than 30 polar expeditions. He died at Framingham, Mass., in 1958. His ashes were taken aboard the American nuclear-submarine *Skate* which dived under the Arctic floes and surfaced through the ice to scatter them in the vicinity of the North Pole.

Chapter 11

The great search for Smith and Ulm

A wave of alarm ran through Australia on April 1, 1929, when radio stations in major cities picked up the message, "About to make forced landing in bad country". The message came from trans-Pacific heroes Charles Kingsford Smith and C. T. P. Ulm, flying non-stop to Wyndham on their way to England to buy planes for their new airline company. The message did not say where they were forced down. Their friends could only guess it was somewhere in the wild and rugged Kimberley country, a sparsely-populated region of dense vegetation and alligator-infested estuaries in Australia's north-west.

From that moment a complete black-out settled over the airmen and their famous plane, the *Southern Cross*. It was as though the untamed land had swallowed them. National anxiety lifted briefly with a message that the plane had landed at a lonely mission station, only to plunge to profounder gloom when the report was declared false. All Australia feared the worst, though the *Southern Cross* was well manned and well equipped. Apart from Smithy and Ulm, it carried their trans-Tasman aides, H. A. Litchfield (navigator) and T. H. McWilliams, who operated the wind-driven radio transmitter and receiving set. The complete silence suggested that the machine had crashed, wrecking the equipment and, perhaps, killing the men, or had overshot the coast and plunged into the sea.

Mystery had fogged the flight from the start. Before he took off, Smithy was said to have received a wire stating that the weather along the route was fine and that the Wyndham landing strip was drying rapidly. He had not long gone before an urgent telegram arrived from Captain Chateau, his Wyndham agent, expressing amazement that the plane had left without his definite O.K., and stating that conditions were unfavourable. The *Southern Cross* should be recalled.

Thus Smithy flew blindly through a "horrible night of torrential rain" into seemingly complete oblivion.

National tension rose as the hours passed. Experts wrote glumly of the 20-foot high vegetation of the Kimberleys which would completely hide the wreckage of a plane. Another reminded his readers that Australia's most ferocious aboriginals who had twice attacked a mission, roamed the wilderness there. He added gloomily that, whereas natives might live off the land, stranded Europeans would starve.

Smithy and Ulm's own airline were the first to act. They telegraphed West Australian Airways who soon had three planes searching the Kimberley area. Pilot Les Holden, barnstorming in his big airliner *Canberra,* was chartered to join the search, while Prime Minister S. M. Bruce's Federal Government ordered Wing Commander L. J. Wackett to get his experimental amphibian aircraft, the *Widgeon,* ready for flight in the shortest possible time.

The first news of the missing airmen came from Pilot J. Woods who flew over the Drysdale Mission Station, 150 miles north-west of Wyndham, three days after the *Southern Cross* vanished. Neither Woods nor the mission had wireless. The only means of communication was by signal. Woods dropped a note. "We are looking for another plane missing since Sunday. Answer these questions in the following manner. Wave sheet to mean "Yes". Place sheet flat on ground to mean "No". First question: Did aeroplane pass here? (*Violent waving of sheet*.) Second question: Did it throw out a letter? (*A native laid a sheet on the ground, while a white man waved another, an unsatisfactory answer of no and yes*.) Third question: Which way did it go? (*A squad of natives formed up and marched south-west*.) The search turned in that direction.

Meanwhile the public became restive. They wanted to know why the Government could not send a service plane from Point Cook instead of waiting three days for the *Widgeon*. While the Government dallied, four Australian heroes might be dying of injuries or starving to death in a tangled wilderness.

Members of Parliament badgered the Prime Minister (Mr. S. M. Bruce) to ask the British to send Southampton flying boats from Singapore to search the bays and islands of Australia's north-west.

Sydney was particularly scornful of official dawdling. The city had no Lord Mayor then. The whole council had been suspended following allegations of graft. It was left to Commissioner John Garlick to call a mass meeting at the Town Hall. "Australia's national heroes are in danger," cried Garlick to the crowd. "They are in dire distress. It is the duty of the Australian people to hurry to their rescue." Within minutes a relief committee was elected and more than £2000 subscribed to a Southern Cross Rescue Fund. Sydney rushed to answer the appeal. The State Government threw in £1000. Teachers, airmen, bookmakers, agents, the staffs of factories and big stores gave generously. Actors stopped performances at theatres to pass round the hat.

Meanwhile, Les Holden was ready with his giant plane the *Canberra* and Wackett with the *Widgeon.* Consternation spread when officials doubted if there was enough petrol for them along the Central route via Broken Hill. Critics flayed the Government for lack of foresight, unaware that oil companies were toiling round the clock to lay down fuel and oil dumps by air, sea and even by camel train. Another howl of rage rose when Holden failed to get off fully loaded from Mascot because the Government had neglected to drain the landing ground and it was nine inches deep in water. Holden had to off-load all his gear, fly light to Richmond and send the gear round by lorry before he could take off.

Wackett was even less fortunate. Half-an-hour after leaving, he was back. It was unfair to ask the *Widgeon* to undertake the task, he said. After six months' idleness in which she had been dismantled and modified, she had not been adequately tested. She was tail-heavy and would not climb above 1000 ft. It would be madness to attempt to cross the Blue Mountains. The *Widgeon* was out of the search.

Again the public turned on the Prime Minister. What was the Federal Government doing to save these men? The answer came promptly. The seaplane carrier *Albatross* was ordered to leave Sydney at once with six seaplanes to search the northern waters. Telegrams recalled officers and ratings who had been given extended leave. Another 8000 gallons of petrol were loaded into *Albatross*, making more than 20,000 gallons in all. On April 12, 12 days after the airmen vanished, *Albatross* ploughed through Sydney Heads and turned north.

By now the search for Kingsford Smith and his mates had built into a major and costly operation. Four planes were combing the Kimberley area; others were on the way. The lugger fleet had left the pearling grounds to search the coast for wreckage. Missionary cutters joined in. Land parties probed into those parts of the Kimberleys where the missing airmen might be. The *Albatross* was belting along at speed 80 miles north of Sydney when the message came that called her back, galvanised the nation and set the brokers cheering on the Stock Exchange. "Found, Found, Found. All Safe."

Les Holden had spotted the *Southern Cross* sitting on the mudflats by the Glenelg River, 20 miles from the Port George Mission and more than 250 miles west of their objective, Wyndham. Four men waved to him. They appeared fit and well. Holden later dropped food, medical stores, mosquito nets, cigarettes and 85 telegrams of congratulation. Next day B. Heath landed his plane beside the *Southern Cross* and returned with the first story of the airmen's 12-day ordeal. Though weak and exhausted, Smithy and Ulm decided to stay with the *Southern Cross* till they could fly her off.

The flight, Smithy revealed, had been a chapter of accidents. An hour out of Richmond their long-wave aerial carried away. This meant they could send messages but not receive them. It explained why they did not get the bad-weather warning and flew on. In torrential rain they overshot Wyndham and were soon hopelessly lost. They came to a huddle of buildings, later identified as the Drysdale Mission, and dropped a note asking the direction of Wyndham. Mission aborigines pointed excitedly south-west. Smithy flew off in that direction thinking he was going towards Wyndham instead of away from it. It transpired later that the natives were pointing to a cleared strip near the mission where the airmen could have landed.

Three hours later, with fuel dangerously low, Smithy and his crew were asking the Port George Mission, farther to the west, how far they were from Wyndham. "Two hundred and fifty miles," came the reply marked in large figures on sheets.

Soon they were nearly out of petrol. Smithy tried to return to the mission, but could not find it. There was nothing for it but a forced landing. They just had time to send out the final message "About to make a forced landing in bad country" before Smithy put the *Southern Cross* down on the flats by a desolate estuary. The wheels ploughed 15 inches into the mud but *Southern Cross* stayed on an even keel. With no wind to drive the generator, their transmitter was dead. They repaired the receiver, however, and were able to listen as the great search swung into action.

Smithy and his aides told of a grim ordeal. When they went for their emergency rations, the locker was empty. Smithy concluded the rations had been stolen. They had only a few sandwiches, a packet of baby food consigned to a Wyndham baby, and a small bottle of glycerine which they used to taste "as it was nice and sweet". They laced their last drop of coffee with brandy, and named the dismal swamp around them Coffee Royal after it. The baby food they turned into gruel and rationed it meticulously. When it ran out, they collected mud snails from the mangrove swamps, boiled them, cracked the shells with rock and forced themselves to swallow the loathsome mess. In a bid to send a message they hitched the transmitter generator to a wheel but failed to raise the power. Horrors plagued them. The creeks swarmed with alligators. They were "bitten almost to pieces" by mosquitoes and flies till they "almost ran blood". They were much too weak to walk overland to the Port George Mission.

Some churches held thanksgiving services for the rescue of Smith and his men. A great shadow, however, hung over the rejoicings. One of the last to set out to search for the *Southern Cross* was the airmen's old friend, Keith Anderson, and his mechanic H. S. (Bobby) Hitchcock. They flew into desert country and vanished. It was as though the desert had swallowed them up.

Chapter 12

He gave his life for his mates

Keith Anderson was confident he would find Kingsford Smith, Ulm and their aides when he set out in April 1929, to search for them. He told the Sydney friends who backed him that he believed the missing men would be found near Port George Mission on the Glenelg Estuary in Australia's wild north-west.

He was in such a hurry to rush to the rescue that he failed to take all due precautions. His plane, *Kookaburra,* carried no radio. The second-hand compass was defective. On the early stages Anderson became lost and had to land and ask the way to Broken Hill. When they landed to make adjustments to the engine, the tool kit was missing. In a wire from Broken Hill, Anderson told his backers he had encountered large magnetic dust storms and his compass went 45 degrees wrong.

Anderson knew the risks he was taking. He had already received a warning from Colonel Horace Brinsmead, Controller of Civil Aviation, advising him to postpone the flight. "This has nothing to do with my search," said Anderson as he read the telegram. "I am flying under private arrangements and intend going on." At Alice Springs Anderson was still confident he would find his friends. "If Smithy not found tonight feel confident *Kookaburra* will pick him up Thursday," he wired. Then, with mechanic Bobby Hitchcock, he climbed into the little monoplane with the air-cooled engine and sped north, intending to make Wyndham in one hop. With them, the airmen took a loaf of bread cut into sandwiches, a pound of cake, some biscuits and three quarts of water. They followed the Overland Telegraph as far as Woodford, 100 miles north of Alice Springs, then vanished in the desert.

The discovery of Smithy and Ulm where Anderson thought they would be, dwarfed for a while the news that *Kookaburra* was overdue. As the hours passed, rejoicings at the rescue turned to gloom. Anderson's backers made a strong appeal to the Sydney *Southern Cross* Committee to send the planes that rescued Smithy to look for *Kookaburra.* "The lives of Anderson and Hitchcock are as valuable as any others," they declared. The citizens' committee, which had collected £7000, directed Les Holden who had found the *Southern Cross,* to switch to the centre in his airliner *Canberra* and look for Anderson. The Federal Government, which had been accused of slackness in the Smithy search, sent five RAAF planes to comb the desert and told the Citizens' Committee they would "stand behind" any efforts to trace the missing fliers. Experts wrote grimly that the task of finding the airmen in featureless, waterless desert was greater and more urgent than the quest for Smithy.

By then, Anderson and Hitchcock had been missing 13 days. The six search planes had criss-crossed 70,000 square miles of desert and one had crash landed. The authorities appealed to Qantas for more help. Qantas Captain Lester Brain left Longreach, Queensland, at once in the airliner *Atalanta.* Eighty miles from Wave Hill, he spotted *Kookaburra,* 200 miles off course, sitting in the scrub with a man prostrate beneath a wing.

A land party of three white men, two natives, a car which quickly broke down, and 26 pack horses left Wave Hill to cut a way through 80 miles of scrub to *Kookaburra.* The party, under Lieut. Eaton, was guided by aircraft which also dropped water and supplies to them. The trek was a nightmare through thick scrub that limited vision to 100 yards. The horses had been 44 hours without water when they reached the crippled plane.

The man under the wing was Hitchcock. Anderson was found dead 400 yards away. He had gone looking for water and had walked in circles till he collapsed. The nearest water was a rapidly drying puddle 45 miles distant.

A pathetic diary on the plane's rudder told of the men's last hours. They had been forced down on April 10 by a faulty push rod which caused cylinder No. 2 to cut out. They repaired the defect then, waterless and in the broiling sun, toiled to clear a strip of bush for a runway. Wheel marks showed they had made five attempts to rise, but the runway was not long enough and they were too weak to lengthen it. Last entry on the rudder diary was made by Anderson on April 12. "No take-off able to be attempted since 11/4/29, due to increased debility from thirst, heat, flies and dust." In their agony the men had drained the alcohol from the faulty compass.

The news of the double death came as a shock to Australia. Wild rumours began to run that the forced landing of the *Southern Cross* was a publicity stunt and that Keith Anderson knew of it. To clear the air, the Prime Minister (S. M. Bruce) appointed a committee of inquiry, consisting of Brig.-Gen. C. L. Wilson, chairman, Capt. Geoffrey Hughes, president of Sydney Aero Club, and Mr. Cecil M. McKay, president of Melbourne Aero Club.

The committee were told of the lost aerial which prevented Smithy from getting the bad-weather message recalling him. They heard of the raging storm which caused the airmen to overshoot Wyndham, of the misdirection at Drysdale Mission and the miracle landing on the Glenelg mud flats at Coffee Royal. Talking of their 12-day ordeal, Smithy said when he saw the rescue plane overhead he burst into tears, he was so weak. Smith and Ulm denied vehemently that they got lost as a publicity stunt and had arranged for Keith Anderson to find them. "That is a deliberate, malicious lie that affects a dead man's reputation," said Smithy heatedly. "It is disgusting. It is disgraceful." Ulm branded it "a deliberate lie, too despicable to talk about". Col. Horace Brinsmead, Controller of Civil Aviation, said it was perfectly ridiculous for anybody to say their forced landing was a stunt, while an expert from the Glenelg area said, "No sane man would stay there unless forced to do so. In view of the nature of the country, I consider it miraculous that they managed even to exist for 12 days."

In its report the committee described Smithy's decision to fly on when he lost his aerial as an "error of judgment". It criticised the failure to make adjustments to the wireless receiving set so they could use it as a transmitter and deplored that they had not checked on their emergency rations before leaving. The report then stated emphatically that there was nothing in the evidence to suggest that the

Australian Harry Hawker . . . a glorious failure in a bid to be the first to fly the Atlantic in 1919.

South Australian, Sir Hubert Wilkins first man to fly from Canada to Europe by the Polar Route. With him is Lady Wilkins.

The first aircraft owned by Qantas, and flown from Sydney to Longreach (Q.), where it was based. J.P. McGinness was the pilot and the date —

forced landing of the *Southern Cross* was premeditated and nothing to impugn the honesty of the crew or any of them. Smithy and Ulm were exonerated.

Meanwhile, as the inquiry ended, another land-party exhumed the bodies of the two airmen from their lonely graves in the desert. Anderson's body was brought to Sydney where, on July 5, 1929, it lay in state at St. Stephen's Church. From early morning to nightfall, thousands filed before the coffin, the queue stretching as far as Phillip Street. Reverently they bowed their heads to the hero whose only thought was to fly to the aid of his mates. Some women knelt by the coffin and prayed. Others sobbed. That evening the coffin was removed to the Presbyterian Church, Mosman, where hundreds more paid homage. Fifty RSL men stood vigil through the night.

Thousands attended the funeral next day. Soldiers marched with arms reversed before the gun carriage, which bore the coffin to the high open space between Sydney Harbour and Middle Harbour where the grave was dug. Five Moths in the form of a cross led eight other planes in a solemn fly-past. Some of the planes came low to drop flowers near the grave. Similar tributes were paid to Hitchcock at Karrakatta, Western Australia. There was consternation later when Mrs. Hitchcock reported that her husband's gold watch and chain were missing from his property sent to her. According to newspaper reports, the watch and chain were found later in a Paddington pawnshop, were redeemed and returned to Mrs. Hitchcock.

Meanwhile, Smithy and Ulm were flying on to greater glory. On July 11, to the acclamation of the world, they touched down near London, having clipped Hinkler's time record of 16 days to 12 days, 21 hours, 18 minutes.

Chapter 13

A girl flies in

It was frankly a battle of the sexes when the English girl flier Amy Johnson set out deliberately to smash Bert Hinkler's solo record of 16 days from England to Australia. She threw down the gauntlet at Bagdad when she was well ahead of Hinkler's time. "When a woman wants to make her mark," she said firmly, "she must do something rather better than any man has ever done it before." Amy ("Call me Johnnie") Johnson did not break Hinkler's record, though she went on to break records in other parts of the world. Her feat, however, sent a wave of feminism flowing through Australia.

Amy Johnson was the daughter of a prosperous wholesale fish merchant and exporter of Hull who, as a young man, took part in the 1898 Klondike gold rush without finding gold. She was a B.A. of Sheffield University, disliked the idea of teaching, so took a job as private secretary with a London firm of solicitors at £3 a week. Then one day, from the top of a bus, she saw planes over Stag Lane aerodrome. That was the day of destiny for Amy Johnson. She joined the London Aero Club, and within weeks had her A licence.

That was not enough for her. She wanted to know how planes were built and why they flew so she talked her way against bitter male resentment into the Stag Lane workshops. For months she worked at the bench from 6 a.m. to 8.30 a.m., went to her job in the city, and returned to the workshop from 5.30 to 10.30 at night, till her father, realising how determined she was, gave her enough money to be a full-time "apprentice". In dirty overalls, Amy Johnson stripped aero engines and built them up again. She learned how to strip and build magnetos and how to recondition tails and wings. Men in the workshop forgot their resentment. They treated her as one of themselves and called her Johnnie.

As she worked, Amy Johnson looked for a challenge to

show, as one paper later put it, "her youthful contempt for mere man". She decided to fly to Australia with her sights on Bert Hinkler's solo record. Like all great air pioneers Amy Johnson suffered intense frustration in her search for financial backing. Smithy and Ulm could approach backers on a man-to-man footing. Amy came in for the "little girl treatment", which infuriated her. Australian Minister for Trade Fenton, then in London, told her, like a kindly old uncle, "Go to Australia by steamer, my girl. You'd be foolish to try to fly there." High Commissioner Sir Granville Ryrie topped this advice by telling her the story of the Red Indian who tried to stop an express train on the prairie. "Everyone admired his courage but not his judgment."

Men aviators laughed at her. The girl was mad to think she could fly to Australia. She had made only four cross-country flights in her life, the longest 150 miles from London to her home town, Hull. She had never flown over sea. She knew absolutely nothing about navigation, they said, which was wrong for Amy had been quietly studying it for months. Newspapers turned her down flat, but Amy struggled doggedly on.

Her first break came when an oil magnate, admiring her persistence, gave her £300 and guaranteed fuel along the route. Her father, who at first thought the plan crazy, gave her another £300. With this Amy bought a second-hand DH Moth with a four-cylinder air-cooled Gypsy engine. The Moth, whose folding wings spanned only 30 feet, had already flown 35,000 miles, some of it in Africa. Amy, then 26, tuned the engine to fighting pitch, painted the craft green and called it *Jason's Quest,* a most appropriate name for a plane bound for Australia, the Land of the Golden Fleece.

Only her father, a dozen airport officials and one press photographer saw her hop into her plane on May 5, 1930, and head for the Channel. She was well prepared for emergency. Piled in the front cockpit were tools, spare parts, tyres, inner tubes, clothes, sun helmet, mosquito net, cooking stove, flints, revolver, medicines, first-aid kit and a long knife to fight off sharks. Strapped to the fuselage was a spare propeller. Experts smiled tolerantly when they heard she'd gone. "The poor girl won't get far," they said. "She can't navigate."

Amy scotched that from the start. Earlier airmen bound for Australia had crossed France and Italy, hopped the Mediterranean and stuck to the all-Red British route. Amy scorned to follow them. Quickest way to India was licketty-split across Europe and Turkey and that was the way she blazed her trail. As a precaution she carried a letter addressed to Turkish brigands telling them not to harm her if she came down and promising ransom. Nearing Bagdad, a blinding sandstorm forced her down in the desert. She had to pile luggage under the wheels to prevent the gale blowing away her plane. Then she stood by, revolver in hand, in case marauding Arabs came. When the storm passed she swung the propeller and was off again.

Amy Johnson arrived at Karachi on the sixth day, slashing Hinkler's record by two days, and was still two days ahead at Calcutta. Then she had her first mishap. She missed Rangoon and came down on a football field 12 miles away, where she ran into a ditch. Fortunately the sports ground belonged to the Government Technical Institute. Engineers from the Institute helped her repair the plane and fix the spare propeller. By now the whole world was taking notice. Newspapers called Amy the "Wonder Girl of the Air". She was somewhat bitter about it. "I had greatest difficulty in getting the small sum needed for the flight," she said. "Now dozens of people are cabling offers of much larger sums for books, lectures, all sorts of things. England only wants to support ready-made success."

In her open cockpit, she survived "desperate peril" over the Java Sea where violent gales forced her to fly a few feet above the water. "I uttered a fervent prayer," she said, "and shortly afterwards saw a double rainbow. I am sure it was somebody looking after me." By now, she admitted, she was "tired, tired, ever so tired." Anxiety came on the leg to Timor when nothing was heard of her for 12 hours. The Dutch were starting searches by plane, ship and land patrol when news came that she had landed near Atamboea and was poised for the final hop to Darwin.

By now all Australia was Amy mad. She had not broken

Hinkler's record but she had put up a grand show. Some became lyrical. One poet apostrophised the elements to protect her over the Timor Sea. One stanza ran:

Skies of Australia kiss her,
Storms hold back your wrath,
Sun of the South throw proudly
Your golden cloaking forth,
Wrap her, ye stars, in glory.
The winged nymph of the North.

On May 24, 19 days out from England, Amy Johnson touched down at Darwin. Five days later 20,000 people at Eagle Farm (Brisbane) groaned with dismay when they saw *Jason's Quest* land, leave the runway and crash through a barbed wire fence. The plane stood on its nose and finished upside down with wings smashed and fabric torn. A great cheer of relief rose when Amy Johnson scrambled smiling and unhurt from the wreckage.

Brisbane folk took her to their hearts but were loath to call her Johnnie. "We as women do not want to call you by a name that suggests manhood or boyhood rather than womanhood," said Mrs. Longman, M.L.A. "The women of Australia are proud that one of your sex has achieved so much."

Amy was too fatigued to fly the repaired *Jason* to Sydney. She was fetched in a Fokker by C. T. P. Ulm and another pilot, Jim Mollison. On the way, Mollison passed her a note, "Will you dance with me at the Flying Corps ball tonight?" Amy replied she would. They did not dance. They were swept apart in the crush. They met again two years later and married — a stormy union that ended in divorce.

Twenty-six planes were in the air as the Fokker approached Mascot. The six planes of the official escort, however, were flown by women pilots, Miss Meg Skelton, Mrs. Bobby Terry, Miss Evelyn Follett, Mrs. A. M. Upfold, Miss Phyllis Arnot and Miss Freda Deaton. The only sour note among the 50,000 who welcomed her came from a Chinese market gardener who was so enraged by fans taking a short cut over his cabbages that he grabbed a double-barrelled shotgun and threatened to blast them to eternity. He patrolled his land shotgun on shoulder till three policemen came to take over the job and prevent bloodshed.

Sydney showered gifts on Amy Johnson. Clubs and famous people rushed to entertain the Queen of the Air, the Mistress of the Clouds and the Lone Dove as she was called. Composers burst into song about her. Orchestras played "Johnnie, our aeroplane girl" whenever she came in sight. Miss Meta Maclean composed "When Amy came flying to Australia," while in England, Horatio Nicholls rhymed "Amy" with "blame me" in "Amy, wonderful Amy, how can you blame me for loving you", which he cabled in code to Sydney so it could be played to her.

The Governor, Sir Philip Game and Lady Game invited Amy to stay at Government House and were somewhat astonished when a horde of admiring businessmen turned their ballroom into a bazaar to offer Amy gifts. They had to have it stopped, and a coolness developed between the Governor and his guest.

Among other things Amy received a sealskin coat, an inlaid glory box, a diamond wristlet watch, a real golden fleece in the form of a sheepskin rug, an opal necklet and earrings, a gold vanity case set with diamonds, pieces of plate and loving cups, a diamond and platinium ring, a gold cigarette case and gold matchbox, a writing desk of Victorian mountain ash, a portable phonograph, a golden boomerang, stockings, hats, tennis dresses, shoes, fountain pens and pencils and tennis racquets. In England among other gifts, £10,000, a new aeroplane and a new car awaited her.

All this caused one Sydney newspaper to suggest, somewhat acidly, that Amy Johnson was only in it for what she could get. Amy burst into tears when she read the story. "This is not fair, not sporting," she said. "All these gifts have been given to me voluntarily." And this was true.

There was an unfortunate incident also, at Canberra. Fog prevented Amy's plane from landing on a Saturday and Prime Minister Scullin had to preside over a complimentary luncheon without the guest of honour. However she arrived next morning and was duly met by the Prime Minister who was roundly rebuked in church by Canon Robertson, for taking part in an official welcome to her at an hour

(11.25) on the Sabbath which showed a total disregard for the feelings of all Christians.

In World War II Amy Johnson became an air ferry pilot. She was taking an RAF transport to its operational base when it crashed into the Thames estuary. Some claim the plane was shot down by an enemy airman. The crew of a trawler saw a parachute open and a figure fall into the water — a tragic end for a very gallant woman. Years later a skeleton, believed to be hers, was found in the estuary.

Chapter 14

The battle for records

In 1929 the age of the record breakers set in. It was sparked by Kingsford Smith and Ulm, who after their tragic mishap in the Kimberleys, set out for London in the *Southern Cross* with Litchfield and McWilliams as crew. They touched down at Croydon on July 11 to world acclaim, having clipped Hinkler's record of 16 days to under 13.

While Ulm returned to Australia to manage their company, Smith set his sights on the first mainland-to-mainland east-west crossing of the Atlantic and the first circumnavigation, though not in one project, of the world at equator level. He took the "old bus" to Fokker's in Amsterdam for an overhaul. June 1930 saw him back in Britain. He flew *Southern Cross* to Portmarnock Beach, Ireland, and, on June 24, took off for Newfoundland, 1,900 miles away. With him were E. Van Dijk, assistant pilot, Patrick Saul, navigator, and John Stannage, wireless operator.

Smithy flew through dense fog so low that twice his trailing aerial dragged through the water. After 31½ hours, he landed at Harbour Grace, Newfoundland, snatched a brief rest and flew on to New York which gave him a hectic ticker-tape parade. The playboy Mayor, Jimmy Walker, pinned medals of honour to his chest, while President Hoover summoned him to Washington for luncheon at the White House.

Smithy and his crew then flew to California where they landed on the airport from which Smith and Ulm had started on their first epic flight over the Pacific. Kingsford Smith, in separate flights, had flown completely round the world.

He was hailed everywhere as the world's greatest airman, the wizard of the air. The newspaper tycoon William Randolph Hearst invited him to his ranch, St. Simeon. Later there was a cruise on the Pacific. Returning to Europe he picked up a new Avro Avian, *Southern Cross Junior,* and flew to Australia in under ten days, setting up a new record and beating several rivals on the way.

To cap this wonderful year, he married charming Mary Powell in Melbourne. Ten thousand people jammed the street to catch a glimpse of the bride. Mounted police were helpless. The crowd broke the barriers, women fainted and pickpockets harvested a bumper crop of wallets and purses.

Kingsford Smith's record of just under ten days for the England— Australia flight was soon under fire. The most serious challenge came from Charles William Anderson Scott, pilot and former boxing champion of the R.A.F., now a Qantas man, tough, athletic, and married to a Melbourne girl. Scott went into strict training for his attack on the record which he made while on leave from Qantas. With his baby daughter's black golliwog as his mascot, he set off from Lympne aerodrome on April 1, 1931, in a Gypsy Moth, powered by a 120 h.p. engine and capable of 110 m.p.h. Mrs. Scott saw him leave.

Scott, making do with very little sleep, reached Darwin on April 10, after a flight of nine days, four hours, eleven minutes, thus clipping more than 17 hours off Smithy's record. He flew on to Sydney where he was escorted in by 20 planes, one piloted by Kingsford Smith.

The capital cities feted Scott. He allowed himself little time for relaxation, however. On May 20, he took off from Wyndham determined to set new records both ways for flights between England and Australia. Suffering badly from cramp and deafness, his face blistered by heat from the engine, he landed in England 10 days 23 hours later, having cut Kingsford Smith's record by nearly two days. Scott now held both records. In recognition of his achievements the King awarded him the Air Force Cross.

Scott's success roused the competitive spirit of one of his main rivals, James Mollison, also ex-R.A.F. and also an ex-amateur boxing champion, small, tough, ready for a fight, who was a crack pilot with Smithy and Ulm's Australian National Airways. Though, when they met, they tried to be friendly, it was no secret that Mollison and Scott did not like each other.

Mollison took off from Wyndham towards the end of July 1931, determined to smash Scott's record to England. He made an emergency landing at Batavia, lost his way over Malaya and his goggles over India — which resulted in very sore eyes, but he flew on doggedly without much sleep.

Mollison was so tired as he flew through fog over France that he decided to land for a rest on the first likely patch as soon as he crossed the English coast. He chose what he thought was a strip of grassland near Pevensy. It turned out to be a beach of shingles which nearly pitched his plane on its nose.

"It was England anyway," he explained later when asked why he didn't fly straight to Croydon. While Mollison rested, nearby residents pushed his plane three-quarters of a mile to better country from which he made a comfortable take-off. Mollison had sliced more than two days off Scott's record.

British pressmen recorded that "a delightful feature of the welcome accorded Mollison at Croydon was Aussie, a famous boxing kangaroo, who led the rush to the airman as the plane landed. His owner grabbed Aussie by the tail just in time to prevent him being struck by the propeller." The rather "correct" Grosvenor House Hotel "looked askance" at the weary Mr. Mollison with his oil begrimed clothes and eyes bloodshot through the loss of his goggles. The wife of a friend rushed out to buy a suit of ready-mades for him, with shoes, shirt, collar, tie and pyjamas before they were appeased. She had a good eye. They fitted.

Not surprisingly, the loss of his records spurred Air-Commodore Charles Kingsford Smith to further efforts to win them back. Flying his Avro Avian and carrying a photograph of the late actress Nellie Stewart as a mascot, he left Sydney for Wyndham, on Australia's west coast in September 1931. The trip was hoodooed from the start. An oil gauge broke. He had to come down at Oodnadatta for repairs and was still doubtful about it when he arrived at Wyndham. Nevertheless he decided to fly on. "Goodbye, see you in 18 days," he said just before he took off, thereby expressing his confidence that he would win back both records without difficulty.

Bad luck dogged Smithy. He had a scare when a gravity tank cut out over the Timor. He experienced bad fainting attacks over the Bay of Bengal. In one of them his plane dropped from 3,000 to 1,000 feet before he recovered, to use his own words, in the "nick of time". For a while he put his head over the side of the plane to benefit from the onrush of air, but was still suffering from severe headache when he reached Calcutta and said he believed he had slight sunstroke. He asked for a glass of brandy, refused to stay for treatment and left an hour later for Jhansi wearing an airman's topee which a member of the aerodrome staff had given him.

A fault in the oil system force-landed Smithy at Bushire (Persia). He arrived unexpectedly at Bagdad after flying through a blinding sandstorm that reduced visibility to zero. Despite his trials, Smithy was nearly a day ahead of Mollison's time. Still determined to break the records, he cabled Henton aerodrome (London) to be ready to overhaul the plane and if necessary install a new engine for the homeward flight.

The watching world was confident he had the record well within his grasp when he took off from Aleppo, Syria, on September 30. Confidence however, turned to alarm when he failed to arrive at Rome. As the hours passed, fears grew that he had been forced down in the wild back country of the Balkans. A day later news came that more fainting attacks had forced Smithy to land near the Turkish town of Milas.

Turkey, at that time, forbade all foreign aircraft to land in her territory. As Smith said, "Police seemed to spring from nowhere and treated me as a suspicious character." For some time they would not even let him cable his friends. "I was kept a virtual prisoner in my room at a hotel," said Smith. "Two sentinels stood outside the door. I was ordered not to communicate with anyone. The police inflicted an interminable interrogation on me. They specially wanted to know why I landed at Milas and not somewhere else." He had to wait till permission came from the Turkish Government at Ankara before he could continue.

Smithy flew on to Athens, still suffering from violent headaches caused, he believed, by sunstroke. He was about to leave for England when he was again seized by faintness. In Athens, he consulted an American nerve specialist. Whatever the advice, he flew on to Rome, experiencing a renewal of weakness on the flight. "The machine did not let me down," said Smithy, sadly. "This is the first time I ever let down my machine."

Though still suffering from headaches and in a highly nervous state, Kingsford Smith was still hopeful of making the return flight and winning back at least one record. "If I get a clean bill of health from my doctor," he said, "I shall rest a few days and begin my flight to Australia in about a week." This was not to be. On October 9, Kingsford Smith left by the liner *Orford* for Australia, hoping to recuperate on the voyage in readiness for another attack on the records. He regretted time had not allowed him to discuss an England-Australia air mail service with Imperial Airways, the British aviation colossus that was already reaching out to many parts of the world.

On reflection, Smithy concluded that carbon monoxide poisoning from a faulty exhaust and not sunstroke had caused the fainting attacks that robbed him of the records and, more important, the chance of early discussions with Imperial Airways on the airmail service.

The battle for records continued. A new thrill came when C. Arthur Butler, a young Lithgow pilot, clipped two hours off C. W. A. Scott's record with a flight of nine days, two hours 29 minutes from England to Australia. The margin was not great but what did thrill the public was that Butler had achieved it in the smallest aircraft ever to fly the route. His single seater Comper-Swift sports monoplane was 5 feet 2 inches high. A man of average height could look over the wing. The hub of the propeller would reach only to his collar. The monoplane's over-all span was 27 feet, its length not much more than eighteen. With wings folded it could fit into a room 19 feet long by 9 feet wide. The seven cylinder radial engine weighed only 130 lb. Top speed was 125 m.p.h.

Butler was born in Birmingham in 1902. He came to Lithgow with his parents 10 years later and in due course was apprenticed to the small arms factory there. He switched to aviation, qualified as a ground engineer, navigator, inspector of aircraft and became a crack pilot holding every class of flying licence. For some time, he barnstormed in New South Wales and Queensland, then joined an aerial services firm.

Arthur Butler, however, was a competent designer. He

planned and built an all-metal, high winged, semi-cantilever monoplane in a shed at Cootamundra. Though his plane was reported to be capable of 185 m.p.h., he received little encouragement in Australia. He, accordingly, packed up and went to England hoping to interest a British firm in his plane or, as an alternative, to become a test pilot.

Britain, however, disappointed Arthur Butler. No one appeared to want his plane and jobs were difficult to get. He was about to pack up and return to Australia when he met Flight Lieut. N. Comper, a brilliant ex-member of the R.A.F. technical department who was building light aeroplanes in Manchester and whose Comper-Swift had come third in the King's Cup air race. Comper liked the young Australian. He let him try his plane and was so impressed by his skill that he agreed that he should fly it to Australia in a bid to break the record.

Butler left Lympne, near London, on October 31, 1931. The plane was so small that he had to squeeze into the cockpit. One eye-witness said it fitted him like a glove. He had one suit and flew in carpet slippers. Weight restrictions allowed him a minimum of food.

Like his predecessors, Butler had his troubles. Some fool tampered with his magneto at Naples; police held him up for a day at Brindisi; he was quarantined for a night at Jask; then he ran into monsoonal rain; Calcutta aerodrome was "under water" when he took off; a thief stole 15 gallons of his petrol at Akyab, Burma.

Despite all this, Arthur Butler broke the record in his midget Comper-Swift. A crowd of more than 10,000 waited at Hargrave Air Park, Sydney, when he flew in, his plane dwarfed by the light aircraft which one newsman wrote formed, by comparison, a "majestic escort".

C. W. A. Scott won back the record the following year when he flew from England to Darwin in eight days, 20 hours, 47 minutes in a DH Moth.

Meanwhile another Australian, Tasmanian-born Harold Gatty, had won world fame navigating the tough, one-eyed American, Wiley Post, on his first record-breaking continuous round-the-world flight. In their Lockheed Vega *Winnie Mae*, Post and Gatty left Roosevelt Field, New York, on June 23,

1931 and flew 15,500 miles over wild territory on the
northern hemisphere route in eight days, 15 hours, 51
minutes, thereby slashing the round-the-world record of 20
days held by the German Graf Zeppelin.

Gatty was born in 1903. At the age of 14, when World War
I raged, he won a cadetship at Jervis Bay Naval College, only
to be turned adrift when peace came and the Government
slashed ships and personnel. However, he was fortunate to
land a junior post in the engine room of a small ship. By then
he was determined to be a navigator. Every free minute of
the night he spent on deck studying the stars. When that job
folded, he had a spell of rabbit trapping in Gippsland, then
took his second mate's ticket and sailed in several ships in the
Pacific trade. Work in Australian waters was chancy. He had
several spells of unemployment and quickly realised that
there was not much opening for a budding expert navigator
in Australia.

Harold Gatty, accordingly, worked his passage to America
where he landed a job skippering a super yacht for a
millionaire. By then Gatty had decided that his knowledge of
navigation would be useful to airmen and had worked out
special methods of applying it. With little capital but
unbounded confidence, he set up an air navigation school in
Los Angeles. Pupils were few and money short. Some airmen
went to him to have their compasses checked. He plotted
courses for others, in some cases he took flying lessons in lieu
of. fees. Gatty plugged doggedly on, and, while running his
school, evolved a revolutionary new ground-speed-and-drift
indicator.

The big break for Gatty came when Anne Morrow
Lindbergh, wife of Captain Charles Lindbergh — the first man
to fly solo across the Atlantic — enrolled at his school of air
navigation. Anne Lindbergh planned to navigate her husband
on some of his flights. She had heard that the handsome,
square-jawed Australian had evolved navigation methods that
were simpler and more efficient than any in use. When she
finished the course, she told American aviation authorities
that the Gatty system was the best in the world.

Gatty's first attempt at a long-distance flight, though it
ended in failure and almost in disaster, also helped his

reputation as a navigator. This somewhat madcap adventure happened in 1930 when, with a tough Canadian, Harold Bromley, Gatty set out from a waterside flying field near Tokyo in a bid to fly non-stop to America, 4,800 miles away across the Pacific. They were not even certain that their monoplane, *City of Tacoma,* carried enough fuel to complete the journey but relied on a tail wind to see them through.

Gatty and Bromley were 1200 miles out when faults developed in the engine and fuel supply system. Though they were not aware of it, deadly carbon monoxide fumes seeped into the cabin and made them ill. They had to turn back. Besides navigating, Gatty had constantly to work a hand pump to keep the fuel supply flowing. By a miracle they made it. The flight, though gallant, had been a flop. The only remarkable feat was that Gatty, flying blind for 2400 miles, had brought the plane unerringly back to the spot where it had first crossed the Japanese coast.

Aided by the Lindberghs, the fame of the Gatty navigation system spread till at last the American Army Air Corps offered Gatty the post of senior navigation engineer in Washington. There Gatty met the tough, one-eyed little adventurer Wiley Post, winner of the Chicago National Air Race. Though friends told him he was mad, Post was convinced he could circle the world by the northern hemisphere route, 15,500 miles, in eight days. All he wanted was a navigator to show him the way. The natural choice was Harold Gatty.

Separated by eight or nine feet of petrol tanks and communicating with each other by tube, Post and Gatty set out from Roosevelt Field, New York, on the morning of June 23, 1931. They touched down at Harbour Grace at 1.17, snatched a hasty meal and rushed back to the flying field to check up on *Winnie Mae* before making the perilous hop across the Atlantic. For the check-up they allowed only three hours, 40 minutes. They were too busy even to pose for photographers and, with what newsmen called "little thought for the weather", took off at 4.57 p.m. for Britain.

"Hello England, we've done it," said Post as they stepped out of *Winnie Mae* at Sealand aerodrome Chester, next day, astonishing the Air Force men who were not expecting them yet. "It's been a splendid trip," said Gatty, "a real joy ride with smooth conditions the whole way except for the first few hours." They had clipped the record for the Atlantic flight established by Alcock and Whitten-Brown by 24 minutes.

Post and Gatty had no time to waste. They refuelled, made a hasty examination of the machine and were off again 20 minutes after landing. At Hanover they were so exhausted from sleeplessness they could hardly speak above a whisper. They forgot to take in petrol and had to return for it. At Templehof, Berlin, Post was hardly able to stand. He gulped down a glass of champagne and was taken to a police station where he collapsed.

Post and Gatty arrived at Moscow on June 26 and were pulled up sharply in their headlong rush by traditional Russian hospitality. They wanted to fly on from Moscow at once. Their first words on landing were, "When do we leave?" The Russians, who a month earlier had quibbled over whether they should be allowed to cross their territory at all, now laid down the red carpet. Blandly and wisely, they opposed the airmen's demand that they be allowed to rush on at once. They had already flown 5,150 miles in three days. Surely they needed some relaxation. Thus Post and Gatty found themselves sitting down that night to a leisurely nine-course banquet with the Soviet authorities, which refreshed them for the testing time ahead. They left at 3.30 next morning.

At Khabaroosh in Siberia, *Winnie Mae* became stuck in a mud hole. Men and horses could not drag her out; they had to get a tractor. Over the Behring Sea the weather was so thick they did not see more than the glass on the cockpit windows. When they climbed high they nearly froze to death. "I thought then I'd rather croak flying across the Behring Sea than crack the ship down south (on the Aleutians) after we were nearly home," said Post when they arrived at Solomon, Alaska. "We managed finally to see enough light to guide the ship here."

Their stay at Solomon nearly ended in disaster. The landing field was rough. Twice *Winnie Mae* tilted nose down trying to take off. On the second attempt, the propeller was bent. Post climbed out with a hammer wrench and pounded

(Above) With the trail already blazed, Captain (now Sir) Alan Cobham (left) set out from London in a De Havilland seaplane in 1926 to survey a route to Australia, accompanied by his mechanic, A.B. Elliott (right). Cobham was the first man to fly to Australia and back.

(Below) Alan Cobham coming down to land his DH50 on the Thames near the Houses of Parliament at the end of his survey flight from England to Australia and back in 1926.

(Above) J.A. (Jim) Mollison after one of his record-breaking flights. Mollison set a new record for the Australia-England flight in 1931, and later married Amy Johnson, but after a few years and a number of joint flights they separated.

(Below) Amy Johnson waving to the crowd after a reception at a Sydney club, following her solo flight to Australia in 1930.

the propeller back into place. To make matters worse Gatty was hit over the heart and on the arm by the propeller while trying to start the engine. He fell to the ground but no bones were broken and, though in pain, he was able to re-enter the machine. At Solomon the airmen were back on American soil, having flown 11,410 miles in six days, 16 hours, 49 minutes.

Post and Gatty flew across Canada and landed at their starting place, Roosevelt Field, New York, where an excited crowd awaited them. The crowd broke through the police cordon and there were many fist fights as police wielding their sticks, forced a way for the fliers to the administrative building. They had taken eight days, 15 hours, 51 minutes to cover the 15,500 miles. Their actual flying time was four days, ten hours, eight minutes.

New York gave the airmen a ticker-tape welcome on July 2. Mayor Jimmy Walker welcomed them personally. Dr. Finley, chairman of the Mayor's Reception Committee, declared that it was particularly fitting that, as an American had navigated Kingsford Smith's *Southern Cross* in the flight across the Pacific, an Australian should have navigated *Winnie Mae* round the world.

In the midst of the ballyhoo, British experts pointed out that, while the flight was a tremendous achievement, it was not round the world but round the northern hemisphere. "The airmen had just taken the top off the loaf." They pointed out that the greatest actual round the world flight was that of Air-Commodore Kingsford Smith, whose trans-Pacific flight remained unequalled.

Gatty and Post wrote a best seller *Around the World in Eight Days.* Gatty sold his ground-speed-and-drift indicator to the U.S. Army and Navy who used it later in developing their famous bomb sight. He went on to distinguished service with the American authorities but never renounced his Australian nationality.

Meanwhile the Australian aviation scene was changing. The commercial age was setting in.

The struggle for the mails

Australia's greatest airmen and their companies were manoeuvring for a glittering prize in 1930. Britain's Imperial Airways had reached as far as Delhi with a regular air service. In quick time it was thought they would be ready to link with Australia. The company to provide that link would rocket to the fore as Australia's premier air-line, with all honour, glory and financial gain such a role would bring. Many young airmen back from World War I had dreamt of such a prize. Now it was within the grasp of those still in the race.

The development of air mail had been a weary process. Ever since the Frenchman Maurice Guillaux carried Australia's first air mail from Melbourne to Sydney in 1914 airmen had waged ceaseless battles with sluggish governments for mail subsidies. They met with constant frustration. Governments were still thinking in terms of railways. They did not want these new-fangled aeroplanes to compete with them.

Postmaster-General William Webster had been particularly unimpressed when young Reginald Lloyd left Sydney with a motorbike convoy early in 1919 to survey a route, somewhat prematurely, for an Australia-England air-line. Said Webster when interviewed: "The whole question of aerial mails is absolutely impracticable as far as this country is concerned. They may be of some value in densely- populated countries, where short journeys are entailed, but here in Australia, with our sparse population and long distances between big mailing centres, the whole position is as different as night is from day. "You said just now," concluded Webster to his interviewer, "that Australia will be the last country to encourage aerial mail services. Let me tell you that, unless I'm very much mistaken, Australia will be the last country in the world to require them." Australian air companies struggled on. They received a financial lift when the Government granted internal subsidies.

Meanwhile the pioneers slashed the flying times between continents, and proved that world air mails were not only possible but imperative. Australian companies competed for the right of linking their country with Britain and the rest of the world. Foremost in the race were Pacific heroes Kingsford Smith and C. T. P. Ulm, spectacular record-breakers and trail blazers, always in the headlines. Against them was an old-established outback line, Queensland *A*nd *N*orthern *T*erritory *A*erial *S*ervices which from joy rides and chartered flights, had built a reputation for solid, reliable, business-like service.

Qantas was born when a car broke its axle and bogged in a Queensland creek in 1920. Driving the car was ex-gunner Fergus McMaster, managing director of a large pastoral firm in Queensland. McMaster trudged back to Cloncurry where he met Paul (Ginty) McGinness, a breezy fighter pilot with serveral kills in the Middle East to his credit. McGinness, and his friend Hudson (now Sir Hudson) Fysh, who also fought in the air over Palestine, had surveyed the route through the Gulf country over which competitors in the first England-Australia air race had flown. They had entered for the race themselves but their backer, Sir Samuel McCaughey, died and the plans fell through. That disappointment led them on to fortune.

When he met the footsore McMaster, McGinness gave up a picnic date with a girl friend to repair the car and send it on its way. McMaster was already air minded. He had first-hand knowledge of the delays and frustration of land-bound transport in the north. He was in a receptive mood, therefore, when Fysh and McGinness met him in Brisbane and told him of their plans for an air company to do joy flights and charter work in the north. McMaster agreed to back them and brought in some of his friends. Queensland and Northern Territory Aerial Services Ltd. was born. The infancy of the company was chequered. Fysh and McGinness had primitive planes to fly; finances were often tight. They had some difficulty in getting suitable machines for the hot northern climate but they struggled through. By 1930 Qantas was a solid company running a reliable service.

By then, too, businessmen in England and Australia were demanding an air mail service which would slash the six weeks or more taken by mails sent by the sea route. Britain's Imperial Airways, though they had reached Delhi, were slow in considering the final link with Australia.. The Australian Government, faced by a looming slump, had its financial worries.

Both were jolted from their somnolence in January 1931, when the Dutch whose air-lines already reached from Amsterdam to Sourabaya, Java, offered to extend their service to Darwin for an Australian subsidy of £26,000. They promised weekly mail planes within six months and announced that they would send an air liner from Java to Australia for a test flight in May.

The offer embarrassed Australian Prime Minister James Scullin who had already consulted the British on the subject but had not got much further. He replied diplomatically that shortage of funds would prevent his ministry from entertaining any offer, British or Dutch, for some time.

They could not cover up for ever, however. Business men in both countries, eager for quicker mails, described the Dutch offer as attractive and cheap. If Imperial Airways could not do the job, why not give the Dutch a try, they asked. Imperial Airways reponded with angry vigour. They worked secretly against time to forestall the Dutch and announced in March that the first official England-Australia experimental air mail would leave Croydon, near London, on April 4, 1931. Letters would arrive in Sydney on April 22, thereby beating those annoying fellows the Dutch by nearly a month.

Britons posted 15,000 letters for Australia by that first air mail. All went well till the last lap when the De Havilland liner *City of Cairo* took over the mail for delivery to Qantas at Darwin. To Australians, it was now all over bar the shouting. Unfortunately, *City of Cairo* encountered head winds which reduced her speed approaching Timor. A few miles from Koepang she ran out of fuel, forcing her pilot to choose what he thought was a nice smooth field for an emergency landing. Unfortunately, there were rocks beneath the grass. *City of Cairo* piled up, wrecked beyond hope of local repair. Her crew of five were only shaken while the mails and films showing the farewell jubilations at Croydon were unharmed.

The crash caused widespread dismay in Australia. Something had to be done quickly, for those persistent Dutchmen held stubbornly to their plans to land air mail from London in Wyndham by May 13 and in Sydney four days later. They also promised to take air mail to London on the return trip, starting from Sydney on May 24. There was no time to lose. Only one man could save face for Britain — Australia's own record setter and breaker Charles Kingsford Smith.

Chartered by Imperial Airways and accompanied by one of his star pilots, George U. (Scotty) Allan, Kingsford Smith set off in his record-breaking plane *Southern Cross* to rescue the mail stranded at Koepang. "It is a great opportunity," he said, "for Australian aviation to show its merits. I earnestly hope that in the near future the air mail authorities will allow Australian aviators to work on the Australian end of the route." Smithy reached Koepang on April 24, picked up the mail and flew it back to Darwin where he landed to the plaudits of the whole town and particularly a crowd of old diggers still celebrating Anzac Day.

There was little rest for Smithy. Waiting for him at Darwin was the first Australian air mail for England, 22,000 letters. He loaded it into the *Southern Cross* and sped off to deliver it to an Imperial Airways plane at Akyab.

While he streaked westward to the admiration of all, his partner, C.T.P. Ulm rammed in another bid for an air mail contract. It was now apparent, said Ulm, that the Australian Government and people would subsidise such a service and make it possible. The occasion demanded that the routes from India to Australia should be operated by an Australian company. Ulm added that his company were working on a proposal for an air mail route between Australia and Calcutta to link up with Imperial Airways. The proposal, with details, would be in the hands of the Federal Government in a few days.

This bid to jump the gun brought a warning growl from Imperial Airways who, through a London spokesman, dismissed the statement as "pious hopes". "Anyone is at liberty to start such a service," said the spokesman loftily, "but it will need a tremendous lot of money." He added warningly that the British Government certainly would not subsidise any enterprise other than by Imperial Airways in that area.

Kingsford Smith picked up the second mail from Britain at Akyab and flew it triumphantly to Darwin. Thanks to him the first experimental air mail had "gone through", but it could hardly be called a triumph of organisation and efficiency. In galling contrast, without fuss, bother or mishap, the Royal Dutch air-liner *Abel Tasman* arrived sedately as promised on May 18 at Sydney where it was welcomed by the Minister for Defence, Ben Chifley, and Lord Mayor Jackson.

The next move in the battle for the air mails came from Kingsford Smith and Ulm. Annoyed by official dawdling, they announced on October 23, 1931, that one of their Australian National Airways machines would leave Melbourne on November 20 with Christmas mail for England. The mail would arrive in London on December 2. The return flight would start on December 10 and land back in Australia with mail just in time for Christmas. There would be space for a limited number of passengers, thus making it the first air-line passenger flight from Australia.

The announcement added that the company hoped by this to convince the Federal Government that there would be no need to call on outside organisations when the Australian section of a regular service to England was considered. The company emphasised that the flight was an ordinary commercial venture and was devoid of any suggestion of stunting or record breaking. Both the Australian and British Governments agreed to send mail by the planes.

Thus it came about that on November 23, 1931, the three-engined Fokker air-liner *Southern Sun* took off from Darwin for Koepang on the first leg of what was announced as the first all-Australian air mail for England. Pilot G.U. (Scotty) Allan was in charge with R.N. Boulton, co-pilot, and

Mr. Callaghan, wireless operator. The sole passenger was Lieut. Colonel H. C. Brinsmead, Controller of Civil Aviation, on his way to London to discuss a regular service. A second passenger, Mr. B. Rubin, had to step down to make way for more mail which numbered 54,530 articles and weighed 1435 pounds.

The first trouble for *Southern Sun* came at Singapore where the wheels sank into the ground, soft after rain. A gang of coolies laid planks on the track and, with the aid of a lorry, hauled the plane to firmer ground. *Southern Sun* was about to take off again when the left wheel sank to the axle nearly wrecking her. Night fell before she was hauled out so they waited for daybreak before continuing.

Next came a shock message from Alor Star. *Southern Sun* had crashed in a paddy field while taking off and was a complete wreck. An early message stated that she was overloaded but this was denied. Crew and passenger escaped without serious injury.

A cable from Col. Brinsmead stated that *Southern Sun* was on the point of rising from Alor Star for Rangoon when an exceptionally glutinous patch of mud slowed her speed from 65 to 50 m.p.h. There was not time for acceleration, turning or stopping, added Brinsmead; the pilot, Mr. Allan, rightly steered the machine over an irrigation dyke and it then sank into a rice field. The machine would have to be completely written off as 10 feet of the wing had been severed and the front end of the fuselage as far as the rear of the passenger cabin had been telescoped.

Several organisations sprang at once to the rescue in support of the traditional edict "the mails must go through". Singapore officials suggested that the Christmas mails be carried in the P. and O. company's steamer *Kashgar* to Colombo, Imperial Airways would then arrange for them to be taken by rail across India to Karachi where they would be picked up by the regular weekly mail plane due in London on December 15. The annoyingly persistent Dutch also offered to fly the Christmas mail to Amsterdam and from there to London. The London Post Office announced somewhat rashly that they would keep faith with the public in the matter of Christmas mail to Australia, using relays of Imperial Airways machines in co-operation with Australian planes.

To Smithy (as everyone in Australia called him) who had staged the Christmas mail service to prove that such outsiders were not needed on the Australian section, this rush to grab his mails was like a red rag to a bull. He left Sydney on November 31 in another tri-engined Fokker, the *Southern Star,* to pick up the stranded mail at Alor Star and take it on to London in good time for Christmas. The Christmas mail would still be an all-Australian affair. Col. Brinsmead, hearing that Smith was on the way, decided to wait and go on with him.

Smithy in turn was unlucky. Heavy clouds and steady rain made visibility poor as he approached Darwin. Flying in to land, his plane struck a telegraph pole at the lower end of the aerodrome. The impact tore the fabric of one wing, broke one rocker arm and one cylinder of the starboard engine, and dented the propeller in two places. The other two engines were undamaged though part of the telegraph wire caught in the port engine. The ground engineer, Mr. Cowper, expressed hopes, however, that they would be able to repair the machine by morning.

Over in Alor Star, Colonel Brinsmead heard of Smithy's mishap. He concluded wrongly that Smith was out of the race and booked a passage in a Dutch air-liner. Smithy, his plane repaired, flew on to Alor Star, picked up the mail and Scotty Allan and flew on. At Bangkok, he was horrified to find Col. Brinsmead in hospital, unconscious and terribly injured. The Dutch plane on which he booked had crashed at Donmuang, 13 miles from Bangkok, due, it was said, to the large amount of Christmas mail it was carrying. Four were killed outright and another fatally injured. Brinsmead was suffering from broken ribs, damaged lungs and contusion of the brain. One side of his body was partially paralysed.

Smithy was terribly upset. He was uncertain what to do till a member of the British Legation gave him the best possible advice. "You can't do anything for Brinsmead by staying here," he said, "You'd better get on with the mail." Smithy flew on and landed the Christmas mail at Croydon on

December 16. "I am very proud to bring the first direct all-Australian air mail to England," he said on arrival. "I hope it will be the forerunner of a regular service. The Australian public has shown that it wants an air mail. If I get a similar load on the homeward trip, it will show that England also appreciates the service."

A message from London added that Air-Commodore Kingsford Smith would spend a busy week consulting the Air Ministry regarding a regular service between England and Australia, for which he considered his machines ideal. The estimated cost of the present flight was £2000 and he was confident a regular fortnightly service would pay for itself in two years if the Government subsidised it at the start.

Luck, however, was still running against Smithy. Scotty Allan was flying the *Southern Star* from Hamble, where it had been reconditioned, to Croydon in readiness for the return flight on December 22, when he ran into mist. He came down to land in what he thought was a suitable field in Kent, only to run into trees which the mist obscured. He and his two companions were only shaken but the machine was seriously damaged. According to one report, the undercarriage was smashed, one engine was half buried, the wings were knocked out of alignment and propellers were damaged. An eye witness recorded that Mr. Allan was making a perfect landing when the plane struck the trees.

Smithy was checked again. He could secure no other plane to transport the more than 70,000 letters that had piled up at the post office for his return trip to Australia. Those diehards, the Dutch, again jumped in with a friendly offer to carry the mails to Batavia if Australia would send planes to fetch them on from there. Smithy reacted with his customary energy. He rushed through repairs to the *Southern Star* and on January 7, 1932, set out with mail for Australia, arriving at Darwin on January 19. He dropped in on Colonel Brinsmead at Bangkok and was able to report he was making some progress.

Smithy's thoughts could not have been happy on that trip. His hopes of winning the Australia-Singapore section of a regular air mail service to England had not matured. He had had informal talks with the Air Ministry but "owing to the absence of Colonel Brinsmead, no concrete plan for such a service could be formulated."

In addition, his firm had lost two planes. One of them, the *Southern Cloud,* had taken off from Mascot for Melbourne in March 1931, with eight people aboard and was not seen again till a bushwalker stumbled on the wreckage and its grim cargo on a rugged slope in the Snowy Mountains in 1958. Now they had lost the *Southern Sun.* The slump was cutting into their revenues. Cash was tight. It looked as though, after all their gallant efforts, they would drift into liquidation.

The only hope for the firm was to snatch a share of the England-Australia regular air service. There seemed some chance in 1932 when C. T. P. Ulm for A.N.A., Hudson Fysh for Qantas and Major N. Brearley of West Australian Airways conferred in a bid to hammer out a combined scheme for the run from Darwin to Singapore. The deciding factor, however, was Britain's mighty Imperial Airways which had already declared that it was essential, "in the interests of efficiency, that Empire air routes should be in one operating hand and in one financial hand." To do this they registered a new company, Australian Empire Airways Ltd., which linked up a few days later with Qantas. Like the duckling which became a swan, the reliable, steady old outback air-line grew into the glamorous Qantas Empire Airways which now sends its jets speeding round the world.

A few days after the Qantas-Imperial link-up, it was announced that Australian National Airways was in process of voluntary liquidation. C. T. P. Ulm, joint managing-director, announced that a new company would be formed to tender for Government air service contracts. Ulm and Major Brearley together submitted tenders for the Singapore-Darwin section of the Australia-England route. The contract went, as expected, to the now-powerful Qantas organisation whose tender was lower, whose resources were greater and whose services were certain to be streamlined and efficient. The visiting Duke of Gloucester inaugurated the Australia-England service at Brisbane in December 1934.

Kingsford Smith was barnstorming in the *Southern Cross* at Canberra when he heard that the King had knighted him. The loss of all they had struggled for came as a heavy blow to

him and to Ulm but they were not the men to repine. Two
more aviation plums remained, the still perilous trans-Tasman
and the trans-Pacific services. They picked up the pieces and
came back fighting.

— Chapter 16

Crashes that made history

The first crushing blow to the hopes of Kingsford Smith and Ulm had come in March 1931, when *Southern Cloud,* one of their fleet of five Avro X air-liners was lost without trace with two crew and six passengers on a flight from Sydney to Melbourne. The aircraft did not carry wireless. In fact, it was stated later that equipment suitable for Australia had not yet been evolved. News of weather along the flying routes was not so detailed as it is today and often did not arrive until planes had taken off. Pilots had to rely on weather forecasts, sometimes on predictions printed in the newspapers.

Saturday, March 21, 1931, dawned wet and squally at Mascot (Sydney). Pilot Travis William (Shorty) Shortridge, 33, R.A.F. trained, a flying instructor and a veteran with 4600 flying hours to his credit, read that heavy rain with thunderstorms lashed most of the air route to Melbourne. He had encountered similar conditions on that unpredictable route many times before, so decided to fly despite the weather.

With Shortridge in the *Southern Cloud* was 23-year-old Charles Dunnell, a pilot-engineer-apprentice, and six passengers — William O'Reilly, 25, a Sydney accountant; Clyde C. Hood, a theatrical producer, speeding to join his actress bride in Melbourne; Hubert A. Farrell, a Melbourne ice-cream manufacturer; Julian Margules, 30, partner in a Melbourne electrical firm and an expert on the new-fangled talking picture machines; and two women, Claire Stokes, an art student, and Elsie May Glasgow, a cook-house-keeper. Theatre-manager Jack Musgrove had booked on the *Southern Cloud* but was too busy to fly that day, while a Melbourne couple decided to extend their holiday in Sydney and switched to a later flight.

Pilot Shortridge took *Southern Cloud* into the air at 8.30

a.m. and headed south against a strong head wind. Though conditions deteriorated badly with gusts of up to 70 miles an hour, no immediate fear was felt when he failed to arrive at Essendon (Melbourne) at 3 p.m. as expected. Average time for the journey was four and three-quarter hours, but some pilots in heavy weather such as this had taken more than seven. By 4.30, however, airport officials were worrying. By then, *Southern Cloud* would be running dangerously short of fuel. They had faith, however, in Captain Shortridge who, in an emergency, would be able to "pancake" the plane safely in any small clearance or even on the tree tops. In that event, they told anxious relatives, the passengers would get a little wet and a little hungry and would, perhaps, have to walk a few miles through bush — but nothing worse.

Attempts to check back over the usual flying course were delayed by the gale which brought down telephone lines in many districts. At last, reports came through that the *Southern Cloud* had been sighted over Albury, nearly two-thirds of the way to Melbourne, while an aircraft refuelling station called Bowser, near Wangaratta, reported that a plane had approached through the cloud and then turned north again as if lost. If that plane was, in fact, the *Southern Cloud,* that placed her even nearer the end of her journey.

Anxiety grew during the night. When Sunday dawned without news there was general alarm. By then it was known that *Southern Cloud* had sped headlong into a cyclone. Pilot G. U. (Scotty) Allan who had flown another plane with five passengers from Melbourne to Sydney on the 21st, spoke of dense cloud going down to 400 feet and a phenomenal wind drift which forced him to lay off 45 degrees to keep his course while flying at 100 m.p.h. Weather men reported thunderstorms, turbulence and strong squally south-west winds, with showers and hail and snow at the higher levels on March 21. There was an almost unbroken sheet of cloud between Sydney and Melbourne, with a depth of 6,000 feet.

News that the *Southern Cloud* was missing touched off one of the greatest searches in aviation history, all of it bedevilled by a mass of conflicting reports. Over the next few days more than 25 planes were in the air searching for the missing liner. Kingsford Smith flew the *Southern Sun,* Ulm his private Avro Avian. Others in the search were J. A. (Jimmy) Mollison, Scotty Allan, Arthur Butler, Les Holden and Fred Haig. They concentrated first on the difficult country between Yea and Kinglake, not far from Melbourne, where several people reported hearing a plane in difficulties. One had even hear a crash and an explosion. There were reports, too, of flares and mysterious flashes, of a bonfire and of a plane circling as if to land.

The searchers concentrated therefore on the heavily-timbered area of razor-back ridges and rocky ravines 40 miles north of Melbourne where there were few houses and no fit landing places. They were hampered constantly by low cloud which made ridge hopping dangerous and plunged the ravine bottoms in deep shadow. Soon the search spread to the Dandenong and Strathbogie Ranges. While the airmen flew, more than a thousand men with packhorses and medical supplies made their way slowly through difficult country to various areas where a plane had been reported. Kingsford Smith caused a minor sensation when the *Southern Sun* sank to her axles in mud while landing at Holbrook, tipped on her nose and smashed a propeller.

As the days passed, more than 500 reports came in concerning the missing plane. No less a person than World War I ace, Squadron-Leader A. H. Cobby reported that he had heard a three-engined plane flying low in the cloud that blanketed Melbourne on the day *Southern Cloud* vanished. This led to the belief that Shortridge, flying blind, may have lost his way, overshot the city and plunged to death in the sea. The theory that *Southern Cloud* crashed in the sea was strengthened when an aero-club pilot fishing at Eildon Weir reported that a low-flying plane had passed south on Saturday at 5 p.m. The coast was searched between Port Phillip and Wilson's Promontory.

Reports of sightings came from such unlikely places as Bathurst in New South Wales and Bega. A party of gold fossickers asserted that they heard a loud explosion while prospecting north of Braidwood, while another reported he had seen a plane in the vicinity of the Snowy where lights had been visible in the direction of Black Mountain. Much time was thus frittered away chasing elusive clues.

Kingsford Smith himself did not believe Shortridge had lost his way as disastrously as many thought. He believed the *Southern Cloud* would be found somewhere in rough country near its usual route. Pursuing this theory, he made three trips to the Snowy and the Alps foothills, one of them over Mount Kosciusko, while R.A.A.F. Wapitis also searched the Snowy area.

After ten days hope was abandoned and searching planes were withdrawn. Pilot Shortridge and his seven companions had vanished without trace in the wilds of New South Wales or Victoria or in the sea.

The Federal Air Accidents Investigation Committee opened an inquiry into the loss of the *Southern Cloud* on April 10, 1931. Mr. C. T. P. Ulm, managing director of A.N.A., assured the committee that every pilot employed by the company was capable of flying blind in any conditions. He agreed that wireless would be desirable in passenger planes and asserted that, if satisfactory equipment could be evolved, his company would certainly install it. The decision whether to fly or not rested in every case with the pilot, said Mr. Ulm. He denied a suggestion that a pilot might take off in bad weather in the belief that if he did not he might be regarded as a "cold-footer". As to health, Shortridge was a man no one ever knew to be sick. "Only trouble I have ever known him have," said Ulm, "was a cold."

Ulm stressed the safety of the three-engined planes. They could fly anywhere on two engines. If two engines failed, which was unlikely, they could fly for 100 miles without landing. If three engines went, an experienced pilot could land on any part of the mountains without injury to passengers. Asked if there were any possibility of a pilot running into a hill, Ulm replied: "It is conceivable."

Pilot James A. Mollison was then asked if there were any areas along the Melbourne route where it would be dangerous to make a forced landing. "There are dangerous places as far as the machine is concerned," he replied, "but not with regard to the passengers. It would be possible to 'pancake' one of those machines and break the undercarriage or a wing but not to injure the passengers."

Chief Engineer F. W. Hewitt told the committee of the daily inspection and regular overhaul of aircraft and engines. He knew of no possible weakness in the *Southern Cloud*. The strength of the wings of that type of plane had been tested by 79 men standing on them. Before each trip, said Mr. Hewitt, the pilot was given a certificate of airworthiness. If a mechanic made one mistake, he was suspended for a month. If he made a second, he was dismissed. Mr. Hewitt discounted the possibility that the machine had been interfered with as a watchman was on duty all night. He added that it might be possible for lightning to destroy an aeroplane.

Col. Brinsmead, Controller of Civil Aviation, expressed the opinion that Pilot Shortridge made over the sea and then came towards Point Cook with the intention of working back to Essendon. He thought the disaster due mainly to the phenomenal weather.

Defence Minister Ben Chifley duly released the committee's report. It entirely exonerated the company and pilot from blame, and declared that the aircraft and engines were airworthy. It expressed the opinion that the extreme weather conditions contributed greatly to the loss but declared that pending discovery of the aircraft, the committee could not definitely assign any cause for it.

The committee, however, recommended that, as soon as possible, the carrying of two-way wireless and a qualified operator should be compulsory in passenger planes, that the official plan for a ground direction-finding organisation be expedited as an urgent measure, that all passenger aircraft carry Verey lights and that a definite code of signals be laid down, that all such aircraft be painted a conspicuous colour to make them easily detected from the air in the event of a forced landing, that weather reports from selected points along air routes be prepared by 7 a.m. to be available to

Wiley Post (top) and Australian Harold Gatty after their historic flight round the world by the northern hemisphere route in eight days, in 1931.

(Above) The Southern Cross photographed from the rescue plane which first spotted it after being lost for 12 days, near the Glenelg River in the Kimberleys in 1929.

(Below) Litchfield, Ulm, Kingsford Smith and McWilliams photographed by the rescue pilot at "Coffee Royal".

pilots at all civil and service aerodromes and that consideration be given to the advisability of carrying a duplicate compass when no wireless navigation aids were available and of fitting a duplicate altimeter where it could be seen by the pilot.

Meanwhile, reports continued to come in about the *Southern Cloud.* Two of them appeared to be cruel hoaxes. In May, a piece of timber was picked up at Seven Mile Beach, Port Kembla, inscribed: "Whoever finds this piece of fuselage torn from the wing of the *Southern Cloud*; we are hopelessly lost. Compass Done, Shorty." In October, a half-caste aboriginal boy reported finding a bottle in Lacepede Bay, South Australia, with a message inside. The message read: "To whom it may concern. This bottle was thrown from the *Southern Cloud.* We are lost and flying about, not knowing where we are. We were over the water when we dropped this bottle." Clairvoyants and spiritualists wrote in with reports of their dreams. In 1947, an aged bushman said he knew where the wreckage of the plane was and was prepared to lead an expedition into virgin bush. Nothing came of the search.

And so time dragged by until, in 1958, 27 years later, Thomas Reginald Sonter, a 21-year-old carpenter, working for a Snowy Mountains construction firm, left Deep Creek Camp, near Cabramurra, for a mountain bush walk on which he proposed to take colour photographs. He was pushing through dense scrub about 150 feet below a mountain ridge top when he ran into a tangle of twisted metal through which trees were growing. A party he took to the scene found aero engines driven deep into the ridgeside and what was left of a telescoped cockpit and cabin. In the wreckage were a number of pathetic relics — a string of beads, three watches, razor, binoculars, scent bottle, shoes, a number of sovereigns, a few calcined bones and a key ring bearing the name of Clyde Hood. Sonter had stumbled across the wreckage of the *Southern Cloud* in the wild country where Kingsford Smith thought it would be.

From the position and the relics, experts were able

to reconstruct the last seconds of the air liner. *Southern Cloud*, they believed, was lost for it was heading north-east in the direction of Sydney when it crashed into the mountain ridge. Though almost on the direct Sydney-Melbourne air route, this was directly opposite to the direction in which it should have been flying. Iced-up windscreen, swirling mist or rain, it was thought, had cut the pilot's vision to a few yards. Shortridge, apparently, saw the looming ridge at the last moment. He pulled the plane into a climb to try to clear the ridge, then gave all engines full power in a starboard bank. This came too late. The plane drove into the ridgeside at speed. The fuselage telescoped. The position of the few bones that remained proved clearly that pilots and passengers were dead before the plane burst into flame.

A similar crash shocked, then thrilled, all Australia in February 1937, when a Stinson air-liner with seven men on board vanished on a flight from Brisbane to Sydney. Again searchers were bedevilled by a multiplicity of reports which led experts to believe the Stinson had crashed in a Hawkesbury gorge or in Broken Bay, just north of Sydney. The search was abandoned after a week.

A young Queensland bushman, Bernard O'Reilly, however, remembering the sudden cyclone that lashed southern Queensland that day, had a hunch that the Stinson might not have got further than the rugged MacPherson Ranges, not far from his home and 400 miles from the area near Sydney on which the searchers concentrated. O'Reilly set off alone to cut his way through jungle mountain country to test his hunch. On the second day, nine days after the plane vanished and two after hope had been officially abandoned, he found the wreckage and beside it two survivors, one seriously injured the other weak from starvation. The coroner who conducted an investigation lashed the people whose false reports had led the searchers so far astray.

The Stinson, a three-engined American monoplane belonging to Airlines of Australia, took off from Archerfield, Brisbane, about 1 p.m. on February 19, 1937. It was piloted by Captain Rex Boyden, who had logged 5000 flying hours, with Beverley Shepherd, 1000 hours, co-pilot. In the cabin were J. Westray, a young London shipping broker, J.S. Proud, a Sydney mining engineer, W. Fountain, a New York architect, J. J. N. Graham, a printers' supplier, and Joseph R. Binstead, a Manly wool broker. The plane, which carried no radio, was scheduled to call at Lismore on its five-hour flight.

Soon after the Stinson took off, a sudden cyclone swept the Queensland-New South Wales border. The plane flew into heavy cloud and battering winds. Despite this, no alarm was felt when it failed to touch down at Lismore. Officials concluded Boyden had flown inland to avoid the storm, had decided to by-pass Lismore, and was flying straight to Sydney. As the hours passed, alarm grew. Telephone checks of towns along the route suggested that the Stinson had been heard over Coff's Harbour, midway on its journey, while another report placed it south of Lismore. Next morning a welter of conflicting reports flowed in. One said the plane had circled a New South Wales coastal town, "emitting smoke". From Maitland came a story of an explosion and of distress rockets soaring into the night. Another spoke of a crash and fire on a mountain.

When the 64 reports of noises and explosions had been sifted and plotted, it was assumed that the Stinson had flown over Newcastle, past Tuggerah Lakes to the mouth of the Hawkesbury at Broken Bay, only 20 miles from Sydney. In the next few days as many as 40 planes were in the air searching the rugged Hawkesbury gorges and the nearby coastal area. An oil patch 100 yards off the coast was taken to confirm the theory that the Stinson had plunged into the sea. Some people even went so far as to say they had seen bodies rolling in the surf but none were recovered. R.A.A.F. planes, alone, covered more than 65,000 miles of country in futile search. Then, after a week, the search was abandoned.

Meanwhile at Lamington, far to the north, young Bernard O'Reilly, who owned a guest house and farm in Queensland National Park, was giving much thought to the missing plane. He knew the Stinson had passed overhead on a direct course for Lismore. He knew, too, that it was battling desperately through cyclonic weather of cloud and gale. He had an idea that, quite likely, it had crashed into one of the nearby mountains and not in the Hawkesbury 400 miles away. O'Reilly got a map and drew a line from where he knew the plane had been sighted to Lismore. The line passed over four mountains cloaked in virtually unexplored jungle. He packed a kit bag with food and set out to search the northern slopes of the mountains to satisfy his hunch.

This was nightmare country. At times he had to hack his way through masses of lawyer vines. He scrambled up slippery slopes, through rocky gorges and spent one night in dripping jungle. On the second morning he looked across a gorge to the fourth mountain on his map. In the vivid green six miles away, he saw a burnt- out smudge and made for it. Eight hours later he hear a faint coo-ee and, hurrying on, found the burnt-out wreckage of the Stinson. Beside it were two men, weak and famished after nine days on the mountainside. Proud's right leg was broken, the bone protruding from the flesh. The wound was fly-infested and gangrenous and the whole leg badly swollen. Binstead was weak from exposure, his hands and legs raw from foraging through thorny undergrowth for berries and water to keep him and his mate alive. It was a strange meeting. "What's the latest Test score?" asked one of them when the excitement had abated. O'Reilly was able to tell them that Bradman was 165 not out in the Test at Lords.

Inside the burnt-out cabin four had perished. Proud and Binstead informed O'Reilly that the Englishman Westray had also survived and had left them to seek help. O'Reilly followed Westray's trail. Westray, he was horrified to find, had fallen over a 20ft. cliff, then crawled half a mile to a creek where O'Reilly found his body, lying propped against a tree. O'Reilly left food for the two men and hurried for help. He returned with a doctor, who saved Proud's leg, and more than 100 bushmen who hacked a path down the mountainside for stretcher-bearers to carry the two to safety.

The crash had happened as O'Reilly guessed. The Stinson had run into foul weather after leaving Brisbane. Pilot Rex Boyden had difficulty in lifting it over the storm-lashed peaks of the MacPhersons. He was within 500 yards of safety when a downward wind tunnel caught the Stinson and hurled it towards the mountainside. Boyden banked to starboard in a last effort to lift the plane, which, however, crashed into a tree, slumped to the ground and burst into flames. Proud, Binstead and Westray were saved because they were sitting on the port side. Despite his broken leg, Proud scrambled through a window and helped Binstead and Westray to safety. The others were trapped in the flames.

The whole country praised O'Reilly's feat. Without his hunch and amazing bushcraft all would have perished. A public subscription raised more than a thousand pounds for him. The coroner exonerated the pilots from blame, but severely criticised those whose false reports had misled the searchers.

Another disaster that made history was the crash of a R.A.A.F. Lockheed-Hudson at Canberra in August, 1940, resulting in the death of three senior Federal Cabinet Ministers, two service chiefs and five others. The Commonwealth then was at war with Germany; Britain's young pilots were locked in a life-and-death struggle with the Luftwaffe in the Battle of Britain. The loss of three Cabinet Ministers and the service chiefs came as a bitter blow to Australia's early war effort.

War news generally was bad when Prime Minister (now Sir) Robert Menzies summoned a special meeting of his War Cabinet. Arrangements were made for Ministers in Melbourne to fly to the capital in a special R.A.A.F. plane. Seats were reserved for Assistant-Treasurer Arthur Fadden and Minister for Customs, Senator Macleay, both of whom cancelled them at the last moment, preferring to go by train so they would have time to confer with their heads of department on the way. Their places in the plane were taken by Chief of General Staff Lt.-Gen. Sir Brudenell White and senior army staff officer Lt.-Col. F. Thornthwaite.

Also travelling in the plane on that fatal 13th of August were Sir Henry Gullett, vice President of the Executive Council, Minister for the Army G. A. Street, and Minister for Air James Fairbairn and his secretary, Mr. R. E. Elford, who were to have flown to Canberra in Mr. Fairbairn's own aircraft but abandoned the plan in favour of the R.A.A.F. plane. The Lockheed-Hudson chosen for the flight was almost new with only seven hours flying time. It was piloted

by Flt. Lt. R. E. Hitchcock, son of pioneer aviator Bobby
Hitchcock, who had as crew Pilot-Officer R. F. Wiesener,
Corporal J. F. Palmer, wireless operator, and Aircraftman C.
J. Crosdale.

The expert ground crew made a thorough check of the
plane before the V.I.P. passengers climbed aboard carrying
brief cases crammed with top-secret official documents and
files connected with the war effort. One of the plane's
engines was rather sluggish in starting. Hitchcock waited in a
corner of the 'drome till all his motors were running
smoothly, then made a faultless take-off. He was in contact
with Melbourne throughout the flight which went smoothly
without a suggestion of alarm. At 10.39 came the normal
message, "I am landing." Watchers at Canberra saw the
under-carriage come down. Some thought Hitchcock was
flying a little lower than usual but there was not hint of
distress.

The aircraft was within two miles of the 'drome when the
port wing dipped sharply and it spun into a nose dive towards
the tree-dotted hills. Hitchcock had almost succeeded in
levelling out his craft when it hit the ground. The wings dug
into the earth and snapped off, the plane hurtled along on its
belly and came to a shuddering halt when it ploughed into a
tree stump. A sheet of flame rose from the crippled plane,
followed by an explosion as the fire touched off the fuel
tanks. Rescuers could not get near because of the intense
heat. Canberra Coroner, Lt.-Col. J. T. H. Goodwin,
subsequently found that all the victims died from fractured
skulls, which meant they were dead before the flames
engulfed them.

Prime Minister Menzies described the crash as a "great
national calamity". Both Houses of the Australian Federal
Parliament and of the British Parliament adjourned as a mark
of mourning and respect. Naturally, there was some talk that
the plane had been sabotaged to check Australia's war effort.
An inquiry, however, decided that the probable cause of the
crash was "the stalling of the aircraft and the consequent loss
of control by the pilot at a height at which it was beyond his
power to regain control". As a result, it was decided that in

future Federal Ministers should not fly in groups of more
than two.

— Chapter 17 —————————————

Women with wings

Meanwhile more women had entered the lists of the record makers and record breakers. The first woman to fly from England to Australia was Mrs. Keith Miller, the wife of a Melbourne journalist. She made the flight with Captain W. Lancaster in the Avro Avian *Red Rose* in 1927-8 two years before the redoubtable Amy Johnson made the first solo flight by a woman in 1930. They made the flight in easy stages to prove the capability of a light aeroplane to carry pilot and passenger over long distances. Captain Lancaster and Mrs Miller left Croydon, London, in October 1927. They received a set-back when they crashed from a height of 50 feet at Muntok on the way from Singapore to Batavia. Captain Lancaster suffered slight concussion and Mrs. Miller a broken nose.

The two fliers landed at Mascot, Sydney, on March 30, 1928, having taken five months on the journey. Their arrival was the closing highlight of the Aero Club pageant. Pageant planes escorted them in. They were welcomed by Governor Sir Dudley de Chair and Lady de Chair, the Minister of Defence, Mr. William Glasgow, the air pioneer, Sir Keith Smith, and a crowd of 100,000 who greeted the little Avro Avian with a roar of cheering. Besides being the first woman to fly to Australia, Mrs. Miller had made the longest flight yet credited to a woman.

In 1932, two years after Amy Johnson's momentous flight, Lady Chaytor and R. Richards flew from England to Darwin in a DH Moth, followed almost immediately by a jolly, friendly German girl, Eli Beinhorn, flying solo in a stock model Klemm monoplane, which as Amy Johnson had done, she serviced herself. Fraulein Beinhorn made no fuss about her trip. She was not out for records or fame. She stated very simply that the main object of her round-the-world flight was to get away from the atrocious European winter. She left Berlin on December 4, 1931, and quickly ran into all the adventure she could wish. A gale forced her down at Aleppo in Syria, another at Bushire, Persia, where she was rescued by the American airman Stephens and the globe-trotting auther Richard Halliburton. She repaid them for their kindness by sewing buttons on their shirts and patching their trousers. She flew alone round Mount Everest and thrilled at the grandeur of a sea of white clouds below huge vistas of snow-capped mountains, with Everest a giant among giants. She added that if one wished to lose all pleasure on that trip it was necessary only to look down on a land of rock and chasm. There was no chance of safety in a forced landing there.

Flying on, Eli Beinhorn met the Siamese Royal Family on the island of Bali and flew into Darwin on March 22, 1932, escorted by R.A.F. flying boats which were on an official visit. She reached Mascot on April 2 to find the whole German colony waiting to shower her with flowers and kisses. One of the first to greet her was Squadron-Leader Kingsford Smith who had last seen her in Berlin. The Mayor of Mascot, Alderman Leveridge was among the notables waiting to greet her. "Miss Beinhorn is not only the first German to fly to Australia," he said admiringly, "she is also the first flier to say she did not do it for the sake of aviation, but simply because she was out to enjoy herself." Fraulein Beinhorn took her plane by sea to Panama and continued her world tour with flights to South America.

An Australian, Mrs. H. B. Bonney of Brisbane, then flew into world headlines. Mrs Bonney recorded her first success by flying round Australia between August 15 and September 27, 1932, in the Gypsy Moth used by Flt.-Lieut. C. W. Hill for a flight from England to Australia in 1930. She had a number of forced landings with minor engine trouble and experienced a shock when her plane and an escorting plane touched in mid-air, bending the rudder of the escort and ripping the fabric on Mrs. Bonney's machine. Fortunately, she was able to fly on. Thanks to masterly handling by the pilot, the escort plane landed safely.

In April 1933, Mrs. Bonney was off again, this time determined to be the first woman to fly from Australia to

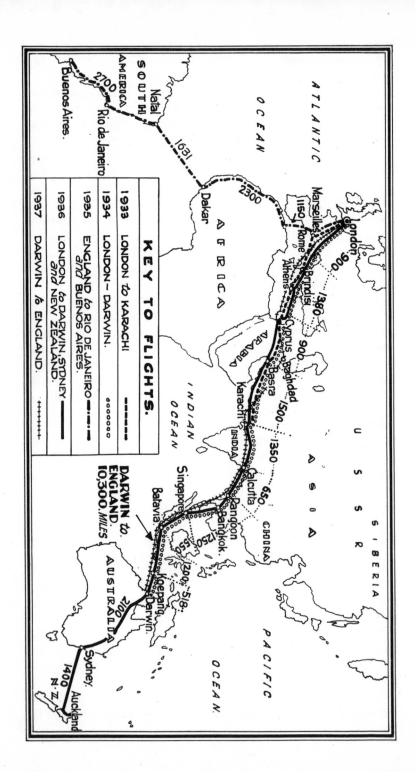

England. She crossed the Timor Sea on April 15 with what was described as a collapsible boat fixed to her plane. Actually it was a large rubber tube with a tin waterbottle in the centre. She was forced down by a terrific storm on the Island of Banbaing off the Siamese Coast and tried to land on a sandy beach. One wheel touched the water, the plane overturned and Mrs. Bonney was thrown out. Again good fortune was with her and she was unhurt. For two days an anxious world waited for news of her while a flying boat searched the islands.

Mrs. Bonney was found by natives. They took a message to the nearest European community who sent a rescue boat to pick her up. She took her plane to Rangoon and then to Calcutta for repairs and flew on. There was anxiety again when bad weather forced her off course over mid-Europe. After some delay she made a perfect landing at Croydon on June 21, thus becoming the first woman to fly from Australia to England. In 1937, carrying a badge of St. Christopher, patron saint of travellers, and a little bronze Buddha as mascots, Mrs. Bonney flew from Australia to Cape Town in easy stages. She encountered storms, had a forced landing or two and recorded that over Africa she flew low over herds of giraffe, buffalo and gazelle.

Meanwhile a Melbourne girl, Miss Freda Thompson, had left London on September 28, 1934, in a Gypsy Moth and, after what she described as a leisurely sight-seeing flight, arrived at Darwin on November 6.

By then, another meteor had flashed across the skies in the person of Jean Batten, a young, attractive "slip of a girl", the daughter of an Auckland (New Zealand) dental surgeon. She had made her first flight with Kingsford Smith over the Blue Mountains and had fallen in love with flying. Jean Batten went to London with her mother and took flying lessons at Stag Lane, where Amy Johnson had learned to fly. She had 130 flying hours to her credit and was one of the few women possessing a B licence for commercial flying when she set her sights on a solo flight from England to Australia.

Jean Batten then became part-owner of a fifth-hand Gypsy Moth, reported to have been owned at one time by the Prince of Wales, later King Edward VIII, and spent months in secret preparation for the flight. Her path to fame, however, was not to be smooth or easy. Only a very determined and confident young woman would have fought the set-backs and frustrations that awaited her.

Jean Batten was only 23 when she left Lympne on April 9, 1933, on her first attempt to fly to Australia. She was caught in one of the worst sandstorms in the Persian Gulf area, was forced down in Baluchistan and, after fitting a new propeller, developed engine trouble and crashed into a hedge near Karachi. She returned to England and in April 1934, quite undismayed, set off again. This time, she lost her way over Italy and, with fuel exhausted, was forced to land in a field near the famous Basilica of St. Paul on the outskirts of Rome. In the process Miss Batten was thrown from the cockpit but, apart from a cut lip and shock, she was uninjured. The machine, however, was badly damaged.

Not even two mishaps could dampen the spirit of Jean Batten. She borrowed a wing from an Italian airman and flew back to England where she had another wing fitted. On May 8, only 16 days after she force-landed at Rome, she set out from Lympne again. Though she announced she was not attempting speed records, she set one by reaching Cyprus two days seven hours out of England. She was delayed by sandstorms and flew through monsoonal clouds so thick she could not see her instrument board and had to use a torch. For a while she flew blind down the Burmese coast. She arrived at Batavia on May 20 with the woman's solo record well in sight. On May 23 she touched down at Darwin, establishing a record of 14 days 22½ hours compared with 19½ days of Amy Johnson.

A fleet of more than a score of aeroplanes met her over the Sydney Harbour Bridge when she flew into Mascot on May 30. Three of the escorting planes were piloted by women — Mrs. Terry, Mrs. Lamb and Miss McKillop. The Postmaster-General, Mr. (later Sir) Archdale Parkhill and the Lord Mayor, Alderman Parker, were among the thousands who greeted her, while the Premier, Sir Bertram Stevens, invited her to be the guest of the State while in Sydney. The intrepid Miss Batten went by sea to New Zealand where she received a rapturous welcome in her home town and was the guest of the Governor-General and Lady Bledisloe.

However, a one-way triumph was not enough for Jean Batten. On April 12, 1935, she took off from Darwin bound for England. Engine trouble forced her down at Foggia in Italy, and bad weather over France robbed her of the record when it appeared to be within her grasp. She landed at Croydon on April 29, the first woman pilot to fly to Australia and back.

Jean Batten now looked to new horizons. She decided she would be the first woman to fly solo from London to West Africa and then across the South Atlantic to Rio de Janeiro and on to Buenos Aires. For this flight she had a Percival Gull aircraft with a range of 2400 miles and a cruising speed of 160 miles an hour. Wearing a New Zealand flag as a muffler and mascot, she took off from Lympne on November 11, 1935. The French authorities insisted that she carry 12 days' emergency rations, two gallons of water and a revolver in case she was forced down in hostile Riff country in Morocco.

She arrived at Dakar, Senegal, in record time, and on November 13, headed west across the South Atlantic. For a while she flew through storm unable to see a yard ahead because of the rain which thundered against her windows. Flying into clear weather, she held out her scarf to trail in the slipstraam in response to the wildly waving crew of a steamer she passed. Jean Batten arrived at Port Natal, Brazil, 13 hours 15 minutes after leaving Dakar, clipping the record for the ocean crossing by more than three hours. Her time from England of 61 hours, 15 minutes lowered the record, held by Jim Mollison, by almost a day. Jean Batten later made a forced landing, due to a leaking fuel tank, 175 miles from Rio de Janeiro but was quickly found by admiring Brazilian pilots.

Miss Batten had one other great ambition. No one, up to that time, had set out to fly from London direct to the land of her birth, New Zealand, making of Australia, the usual terminus, just a stage of the journey. On October 5, 1936, she set out from Lympne to do just that, "hoping to halve my Australian record" on the way. Flying by night and day, frequently by dead reckoning through storms, she arrived at

Darwin on October 11 after an astonishing flight of five days, 21 hours, cutting the previous solo record set by H. F. Broadbent by 24 hours. In performing this feat, she displayed amazing stamina. Until she reached Koepang she had had only seven hours sleep in five days. At Koepang she snatched only six before setting out on the perilous fight across the Timor Sea. At Koepang, too, she thought she had been robbed of success. The inner tube of the tail wheel burst while her plane was being wheeled over rough stones to the fence. "I felt my chance of the record had gone," she said later. "I was so disappointed I nearly cried." A friend, however, bought up all the sponges in Koepang and stuffed them into the wheel, making it sufficiently firm for her to take off for Darwin.

Jean Batten had made the flight in seven "hops" — Brindisi, 1309 miles; Basra, 1760 miles; Karachi, 1341 miles; Akyab, 1761 miles; Singapore, 1594 miles; Koepang, 1647 miles and Darwin, 511 miles.

Now came the last stage of her flight — over the stormy, unpredictable Tasman to New Zealand. Friends and experts tried to persuade her not to attempt it. She had made a magnificent flight from Britain. Surely she would be content with that and rest on her laurels. The mayor of her home town, Auckland, cabled urging her not to make the Tasman flight. The Australian Controller General of Civil Aviation said he hoped Miss Batten could be persuaded not to make the attempt.

Having had her plane overhauled, Jean Batten was adamant. "Those who take on flying cannot be frightened," she said. "I spent six months organising this flight and I am confident of seeing it through to my homeland. I do not embark on long-distance flights without weighing everything. This Tasman flight is not being undertaken rashly." On October 16, Jean Batten flew the Tasman setting a record of nine hours 29 minutes for the crossing and a "first" record for the England-New Zealand flight of 11 days 56 minutes. She was the first pilot, man or woman, to make a direct flight from England to New Zealand and the first woman to fly solo across the Tasman. New Zealanders gave her an unforgettable ovation wherever she went.

In 1937, Jean Batten made another attempt on the Australian-England solo record. This time she made no mistake. She landed at Lympne on October 24, having completed the flight in five days, 18 hours 15 minutes, cutting the record held by H. F. Broadbent by more than 14 hours.

Chapter 18

Flight to Hell

Meanwhile, as Eli Beinhorn arrived triumphantly at Darwin in March 1932, two other Germans, Captain Hans Bertram, and his mechanic, Adolph Klausmann, were on their way to Australia in what they described later as a "Flight to Hell". Their adventures were another grim warning to airmen inclined to tangle too freely with the great Australian wilderness. Bertram and Klausmann left Cologne, Germany, in February 1932, in the Junkers *Atlantis* bound for Australia on a goodwill tour seeking new markets for Junkers aircraft products. They were good airmen, their craft was good and, as expected, they arrived without incident at Timor.

Bertram had no qualms about crossing the Timor Sea. The Junkers carried fuel for 7½ hours flying and should complete the trip in five which would more than cover any ordinary risk. They set out soon after midnight on May 14 and were so confident of an easy landfall that they took only two bananas to eat on the way. They reckoned without a huge cloudbank which straddled the Timor Sea. They flew low, hoping to fly beneath it, but found it went right down to sea level. Then they climbed in a bid to get above but found it just as thick at 9,600 feet where they nearly froze in their tropical clothing. Bertram was forced to descend to a warmer level and fly by compass through the dense mist.

Long after the five hours had passed, they still droned on. Their fuel was getting low and Bertram was more than anxious when, at last, they broke from the cloudbank into clear weather and saw a stretch of coastline ahead. Bertram set the Junkers smoothly down in a sheltered inlet and taxied to the shore. He and Klausmann solemnly shook hands when they landed. They thought their flight was successfully over. They could not have been more wrong. Fifty-three days of scorching horror lay before them, 53 days of starvation and

Guy Menzies of Sydney, in shirt sleeves, landed on deceptive marshland when he finished the first solo flight across the Tasman from Australia to New Zealand in 1931.

(Above) Mrs. H. Bonney, of Queensland, the first woman to fly solo from Australia to England. She made the trip in a DH Moth in 1933.

(Below) One of the greatest of them all, the New Zealander, Miss Jean Batten, who set solo records between England and Australia and New Zealand, and flew the South Atlantic to Rio and Buenos Aires. She made the flights between 1933 and 1937.

privation which led Bertram to record his mission under the title *"Flight to Hell"*

Though his fuel tanks were almost empty, Bertram had no fears that first day. He believed he was on the north coast of Melville Island and, if help did not arrive within hours, had only to trudge across the island to the settlement where, no doubt, he would be able to obtain fuel for the short trip across the narrow Clarence Strait to Darwin. He was so confident that when night fell they slung their hammocks and slept peacefully — the last untroubled sleep they were to enjoy for many a long day.

Bertram and Klausmann were blissfully unaware that they had drifted far off course in that cloudbank. They had passed 250 miles west of Darwin and had penetrated 150 miles down the Joseph Bonaparte Gulf to make their landfall in wild country near Cape Rulhieres between Wyndham and the Drysdale Mission. Like Smithy, Ulm and their party before them, they were marooned on the desolate Kimberley coast where no white men lived and where some of the few natives were believed to be cannibals.

They woke refreshed next morning to find a naked aborigine staring at them from the rocks. Bertram hailed him cheerfully and asked him in English where the nearest fuelling station was. The native grinned and took off into the bush. Bertram did not mind. He firmly believed they would find a settlement not far along the coast, possibly round the next headland. A short flight, however, revealed only an arid coastal wilderness. They came down in a sheltered creek, their petrol tanks bone dry. Already they were parched. They searched for fresh water but found none. Bertram and Klausmann filled a water bag from the aircraft's radiator, bundled up their razors, toothbrushes and other things they thought necessary and set off eastwards along the coast, expecting to find a settlement any minute. A sharp stone gashed the waterbag and spilt its contents on the sand. Three days later, famished, tortured by thirst and footsore they were still stumbling on.

They stripped to cross a creek with their bundled clothing on their heads and were making good headway when they saw three crocodiles slide from a sandbank and make for them. In their wild scramble to shore they lost their clothing and their boots. All they had was their sun helmets and Bertram's pistol. Naked, they set off back to their aircraft. They covered themselves with sand to sleep that night but were still tormented by mosquitoes.

A noise in the bush woke Bertram next morning. Standing a few paces away was a fully-grown kangaroo whose blood and flesh might have meant life or death to the two stranded airmen. Bertram took a shot at it with his pistol, missed and was distressed to see it bound off into the scrub. Bertram now came close to despair. He guessed he had been hopelessly out in his calculations. For three days he had been without food or water; his feet were bleeding; an agonising death might lay in wait for them. The pistol offered an easy way out but he was too much of a fighter to take it. To avoid any temptation in the troubles ahead, he rose to his feet and threw the pistol far out to sea.

Four days later, burned by the sun, tortured by mosquitoes and almost raving from thirst they arrived back at their aircraft where the last few pints of water in the Junkers' cooling system saved their lives. They put on spare clothes and rested. They thought themselves lucky when Klausmann found a rusty fishhook in his tool kit. With this they caught a small fish which they turned into soup, but lost their hook on the next throw. Mercifully a shower broke over the coast that night, enabling them to fill all their containers with the water running from the plane's wings.

Meanwhile a massive search had been launched in northern Australia for the missing men. Launches, aircraft and police patrols searched the coast west of Darwin as far as Wyndham where the searchers stopped. This limitation was reasonable for no one would credit that an airman could get 250 miles off-course in a 500-mile flight. Hope was abandoned. The German Consul in Sydney cabled Berlin that the search was off and the men should be considered dead.

He was wrong. Hans Bertram was of the stuff that makes heroes. Though hungry and weak, he set to work to detach one of the floats from the Junkers and turn it into a boat. Next they fashioned rough paddles, fitted a mast to it, and made sails of bathrobes and a pair of trousers. In this frail craft, Bertram and Klausmann put to sea in hope of intercepting a passing ship. The float was almost uncontrollable but somehow, aided by tides, they managed to coax it over seas that threatened to swamp it. On the third day they sighted a cargo ship. Bertram fired red flares from a signal pistol they had retrieved from the plane. They waved and shouted — only to see the ship chug by less than half a mile away without sighting them.

Klausmann broke down and sobbed. Bertram handed him his paddle and they set off to row back to land. Two days later they stumbled ashore and made their way with difficulty back to their aircraft. By now they were out of water and, apart from a few shell-fish, they had not eaten for 19 days. Klausmann had periods of delirium in which he babbled of bakers' shops and choice cuts of meat at the butchers in his home town in Germany. Battling grimly for survival, chewing the leaves of trees as they went, they wandered inland and stumbled across a rock pool of fresh water. They spent the rest of the day soaking and drinking in the pool.

Somewhat refreshed, they decided to push southward to test Bertram's original theory that they were on the north coast of Melville Island. Bertram set their course by the plane's compass as, for three days, they trudged and stumbled through thick scrub and over stony ridges which gashed their hands and knees. They gave up and returned northwards when, from a rise, they saw mile after mile of grey Australian bush spreading, it seemed, into infinity. They were so weak they had to rest every 30 yards on their trek back to the coast and when at last they reached it were horrified to find they had missed their waterhole. Bertram took another two days to find it and, when he returned with the news, found Klausmann crawling up the beach shouting hysterically.

By now they realised they were on the north-west coast of the mainland and not on Melville Island. They decided not to go far from their waterhole. They found a cave nearby, made beds of grass and ventured out only for water and to search for shellfish and lizards which they ate. A plane flew over but the pilot did not see their frantic waving or the red light from the signal pistol.

As if their trials were not enough, Bertram was now tortured with toothache. Unable to bear the pain he ordered Klausmann to take the tooth out. Klausmann reluctantly obliged. With Bertram's head on his knees, he tugged at the abscessed tooth with pliers from his tool kit, broke it, pierced the abscess with a safety pin and took out the rest of the tooth piecemeal.

Meanwhile, the stars in their courses were working miracles for the lost men. Father Cubero of the Drysdale Mission, on his way back in the Mission launch from a repair trip to Wyndham, anchored in a lonely creek where a native showed him a cigarette case engrave H.B. The case had been in the clothes Bertram dropped when the crocodiles chased him from a creek. It was a chance in a million that it was found. Father Cubero realised at once that it belonged to one of the missing airmen. He sent a native runner to alert the Wyndham police to a new search and sped himself to Drysdale where he turned out most of the mission natives to seek the missing men.

Constable Gordon Marshall set out at once from Wyndham with 50 horses and mules loaded with food and medical supplies, bound for the creek where the cigarette case had been found. It was a race between Marshall and the native mission boys and the aborigines won. Hunting as they went, they reached the region of the cave within a few days. One was actually spearing fish 100 yards away when Bertram and Klausmann saw him as they staggered from the cave on the 37th day from their landing. They shouted, whereupon the aborigine leapt across the rocks with a fish which they devoured ravenously. They were amazed when the native dropped to his knees in their cave and offered a prayer of thanksgiving.

Smoke signals brought other aborigines from the mission with tinned meat and flour and when that was gone, some of their deliverers stayed to hunt for them. When the kangaroo meat they brought was too tough for the weakened men, they kindly chewed it for them first.

Marshall's party made slow progress through the dense

scrub. They took 14 days to cover the first 100 miles, spurred on always as the bush telegraph brought rumours that the white men had been murdered. A wild native guided Marshall to where he had seen the tracks of the whites. There Marshall encountered one of the mission natives who had stayed with the castaways to hunt.

Marshall pressed ahead alone with food and medical supplies to be greeted at the cave by two emaciated men who babbled mainly of bread. He gave them a sip of brandy, bathed their wounds and fed them sparingly on softened food till they were fit to travel. Bertram's stamina and spirit pulled him through. Klausmann was not so well. His overstrained nerves snapped. For some time he was delirious and had to be strapped and watched closely on the journey to Wyndham, which they reached 53 days after landing. Klausmann went home to Germany by sea. Bertram salvaged his seaplane, flew it on a goodwill mission round Australia and then back to Germany, accompanied on part of the journey by A.N.A.'s crack pilot G. U. (Scotty) Allan and a stowaway.

Meanwhile another great aviation event loomed on the horizon, the Melbourne Centenary Air Race. This was to bring more strife to Kingsford Smith.

— *Chapter 19* —————————————————

The great race

All Australia thrilled on October 23, 1934, when a scarlet racing plane streaked above a finishing line at Flemington racecourse, Melbourne, and touched down at Laverton only two days 23 hours after leaving England. The pilots, ex-Qantas man C. W. A. Scott and T. Campbell Black, world-war air ace, coffee planter and aerial crop duster, whose mother was Australian, had clipped the record for the 11,323-mile journey by more than four days. They had also won the £10,000 first prize in the Melbourne Centenary Air Race.

While they received the plaudits of the crowd, a great Australian, Sir Charles Kingsford Smith, who a few weeks earlier had been sent a white feather, was making the first flight from Australia across the Pacific to the United States. By selling his plane in America, Kingsford Smith hoped to repay his backers for his failure, through sheer mischance, to carry the Australian flag — they hoped to victory — in the Centenary race.

The race had a chequered history. Sponsored with £15,000 prize money by Sir MacPherson Robertson, the Melbourne chocolate millionaire, it roused the enthusiasm of the world. Americans, Italians, Germans, Dutch rushed to enter. The Americans declared they would enter a fleet of super machines capable of the then phenomenal speed of 250 miles an hour, led by their ace speedster Clyde Pangborn. Britain's De Havilland Company, already famous for record-breaking planes, set to work to design and build the famous Comets in which they were sure such pilots as Scott and Black, the redoubtable Mollisons and veteran Cathcart Jones could whip the world. Wiley Post, the round-the-world flyer, entered with his record-setting plane *Winnie Mae*. Australia's Harold Gatty intended to fly, while from France came the news that the first cross-Channel airman Louis Bleriot, then over 60, was building a plane with a retractable under-carriage, capable of 217 miles an hour, specially for the race. Another early entry was Harry Lyon, who had navigated Kingsford Smith and Ulm on their great Pacific flight from Oakland, California, to Brisbane.

Naturally Australians hoped an Australian would win. Some hoped Australians would win in an all-Australian plane which, powered by two Harkness Hornet engines, was soon under construction at Mascot. It was to be flown by Mr. D. T. Saville with Mr. L. J. R. Jones, who was part designer of the plane and who was lecturer in aircraft construction at Sydney University, as co-pilot. The plane was expected to reach 250 miles an hour.

Australians pinned their main hope, however, in their greatest airman, Sir Charles Kingsford Smith, the first man, with Ulm, to fly the Pacific, the first to circle the globe, and who had flown several times between Australia and England in the see-saw race for records. An anonymous patron, afterwards revealed as Sir MacPherson Robertson himself, and other friends guaranteed Smithy £5000 to help finance the Australian challenge. There was some regret when Smithy declared he would prefer an American plane. The Americans, he explained, because of their great distances, had developed high-powered, long-distance machines, whereas the British, owing to the smallness of their island and consequent restricted distances, lagged behind in this regard.

Sir MacPherson Robertson insisted, however, that he would prefer Smithy to fly a British aircraft. Smithy agreed if one could be found. They cabled the De Havilland Company asking if they could build a fourth Comet for Kingsford Smith. De Havilland said they could, but later it transpired, the plane would not have a variable-pitch propeller which would cut Smithy's range by 400 miles compared with other competitors. With such a handicap, Kingsford Smith would have little chance of victory. Reluctantly Sir MacPherson Robertson agreed that he should buy an American machine. To the grumbles of a soured few that this was "un-British," Kingsford Smith went to America to buy one.

Meanwhile the race was running into strife. The Americans complained that safety conditions laid down for the race

were slanted to favour the De Havilland Comets and to put
the Americans out of the running. The conditions were that
planes should conform substantially with minimum requirements for air worthiness laid down by the International
Commission for Air Navigation to which the Americans
steadfastly refused to belong. The American National
Aeronautic Association protested to the British Aero Club
against the conditions. The Centenary Air Race Committee
refused to alter them. They had been adopted as being the
only international safe standard for the race, they said, and
they did not intend to create a precedent by allowing one
nation to lower the standard. The conditions set out, they
added, did not debar any well-known type of aircraft in the
world from competing if they were not overloaded.

The wrangle provoked such hostility in America that Mr.
Dow, Australia's representative there, cabled Prime Minister
Lyons that the impression was growing in the United States
that the rules for the Centenary Race had been designed to
exclude American aeroplanes and suggested that the Prime
Minister make an announcement. Mr. Lyons side-stepped the
issue. Airworthiness conditions were a matter for the Air
Race Committee, he replied, adding, in effect, that the
Government wouldn't buy into any such quarrel. The
Centenary Air Race Committee refused to alter the
conditions, but said they would leave approval of American
aeroplanes to the supreme American authority for aviation,
the Department of Commerce, who would be responsible for
seeing that all American entries met the conditions. "Any
change in conditions would lower the margin of safety," they
repeated, adding, somewhat tactlessly, that a repetition of
the tragic Dole race from San Francisco to Hawaii in 1927,
when seven died, would be deplorable. Sir MacPherson
Robertson stressed this angle, saying that, though there was
no intention to debar anyone, no advantage would accrue to
the cause of aviation by making the race one for freak
machines. Thus it was left to the American Department of
Commerce to judge American planes with the strict
understanding that they would have to stand up to rigid final
tests in England before the were allowed to start.

The Americans continued to be suspicious. This was

demonstrated by Colonel Roscoe Turner, one of their ace pilots, who complained bitterly and prematurely in July 1934, that his entry for the air race had been rejected by the Royal Aero Club on technical grounds because he made a mistake in cabling the designation of his craft. The mistake was that he had entered a Douglas and later changed his mind and said it was a Boeing. For some reason a reply came that he must adhere to his original entry.

Without further inquiry Col. Roscoe Turner launched a bitter denunciation against the Royal Aero Club, who had agreed to supervise the race. "It was my understanding," said the gallant colonel, "that the Robertson race was a sporting event for the purpose of building up international goodwill, but the Royal Aero Club seems to think differently. It placed as many technicalities on the race as it possibly could. Before the American pilots could find out what kind of planes were eligible, the entries were ready to be closed. I cabled my money and entry, but because the specifications and details do not conform with my cable, I have been disqualified. But," he added menacingly, "I have plenty of air in America to stir up. It has cost every American pilot from 35,000 to 100,000 dollars depending on the type of plane, to enter the race, with the possibility to win a 50,000-dollar prize," Col. Turner proceeded, "so it is easy to see that they have not entered from the monetary standpoint. I feel that the Royal Aero Club, instead of trying to help American entries to get into the race, is doing everything it can to keep them out. As there are always technicalities to tie up things with an aeroplane if you look for them, all the American entries will have plenty of trouble ahead regardless of how good they are."

Colonel Turner spoke too quickly. He had hardly voiced his protest before a cable arrived from the Royal Aero Club announcing blandly that his entry for the Melbourne Centenary Air Race with a Boeing plane had been accepted. "I would like to retract the statement I made yesterday," said Turner. "I also apologise to the Club for being a bit hasty but it was due to my great disappointment. All my associations with English fliers have been most pleasant. I have found them sportsmen in the highest degree."

On July 18, 1934, it was announced that 64 had entered
the race, though one had already died, unfortunately, in an
air crash. Numbers allotted to Australians included Sir
Charles Kingsford Smith, 28; Mr. Ray Parer, hero of the
aerial jalopy saga in 1919-20, 35; Mrs. Keith Miller, 36; Mr.
D. Saville, 43; and Mr. Warren Penny, 64.

By then the spotlight had turned firmly on Sir Charles
Kingsford Smith on his way back to Sydney in the *Mariposa*
with his American single-engined Lockheed Altair monoplane
housed on the upper tennis deck. The sleek plane with the
silver wings was said, by one report, to have cost 21,000
dollars. Across the dull blue fuselage broad silver arrows and
the proud name *Anzac* were painted. "I am already in strict
training for the race," said Smithy. "I am on a diet and have
cut out cigarettes and liquor." When *Mariposa* docked,
crowds of admirers rushed to the upper deck to see the plane
which they hoped would win the prize for Australia. All,
however, were not pleased with Smithy. Many still thought
him un-British for not flying a British plane. The tide of
hostility mounted.

To cap it, Smithy ran into a maze of red tape. Customs
officials refused to let him offload the plane till they had
covered with brown paper the name *Anzac*. The name was
protected in Australia and, though he was an Anzac himself,
he was not allowed to use it. Smithy had the bright idea they
might let him take the plane from the ship at East Circular
Quay to Macquarie Street and take off from there for the
flight to Mascot. This, understandably, was banned.
Accordingly, he had the plane slung by floating crane to a
lighter which took it to Anderson Park, Neutral Bay.

Several thousand people saw Kingsford Smith and his
co-pilot, P. G. (later Sir Gordon) Taylor, take off there for
Mascot, where officials, wrapped in red tape, removed the
high-tension leads and solemnly affixed seals in their place.
On instructions, Smithy painted out the name *Anzac* and
re-christened the plane *Lady Southern Cross* in honour of his
wife.

Smithy continued to flounder in a sea of red tape and
rising criticism. When the Customs Department released the

plane, he found he could fly it only within a radius of three miles of Mascot, pending a decision by the Department of Civil Aviation in Melbourne. Questioned about this, the Director of Civil Aviation, E. C. Johnston, replied that no application had reached the Department for the registration of Sir Charles Kingsford Smith's Lockheed Altair. Until registration was granted the aeroplane could fly only for experimental or testing purposes within three miles of an aerodrome. On registration it could be flown anywhere.

Meanwhile Smithy had replied to his critics. He denied that he was un-British in buying an American plane. "I am British to the backbone and proud of the fact," he said. He proceeded to explain that British manufacturers, because of the geographical limits of their island, had not yet had time to develop an aircraft of the type required in competition with American manufacturers who had been building such planes for years. He told his critics how he had tried to obtain one of the so-far-untried Comets, only to be told he could not have a variable-pitch propeller, which would have put him out of the race. "Probably the biggest percentage of my critics," he said, "are owners of American cars, American refrigerators, American wireless sets. Surely I am the best judge of what I need for the race. I consider I have a better chance of putting up a good show in the race with the aircraft I have selected. If I had been forced to use a plane of which I did not approve and came in a bad last, I would get no sympathy." He regretted the hold-up in licensing the Altair and wished the department could administer the law more in the spirit than the letter.

The Lockheed Altair *Lady Southern Cross* was registered on July 30, 13 days after *Mariposa* docked. Kingsford Smith and P. G. Taylor promptly tested her and proved her speed by breaking most of the records for flights between Australian capital cities.

As the weeks passed Smithy ran into more trouble. The first indication of a serious setback came with the announcement that "because Sir Charles Kingsford Smith failed to obtain a certificate of inspection by the American Department of Commerce of his Lockheed Altair plane, there is some possibility that the machine may not be admitted to

England." The fault, it transpired, was entirely Smithy's. Before he left for the United States, Controller of Civil Aviation Johnston had telephoned to warn him that he should obtain a certificate from the United States authorities that the machine conformed with the normal requirements of the international convention and that he should also obtain from the American authorities a certificate of airworthiness. This Smithy, in his hurry, neglected to do. Without them he would not be able to compete.

The Australian authorities did their best to help him. The British Embassy asked the American Department of Commerce for the necessary documents. The Department replied they would have to wait for data from the Lockheed factory which information, for some reason, was slow in coming. The Embassy pressed again, without result. "There is some fear," came the message, "that the Department of Commerce may contend that it cannot issue a certificate for an aeroplane which it has not seen and which was shipped from the country before it could be inspected for details on which admittance to England apparently depends."

The weeks passed. Famous firms in many parts of the world worked against time to get their entries ready for the start of the great race on October 20, 1934. Propped up in bed suffering from a chill in Melbourne, Kingsford Smith said on August 30 that he had not anticipated the irritable hitches and obstruction that had occurred arising from his failure to obtain a certificate of inspection for the plane from the American Department of Commerce. He made no criticism, but he did expect that a helping hand would have been more in evidence. "Time is running short," he said, "and it will be necessary for me to leave within a fortnight or so if I am to compete in the race. If the certificate does not arrive from America, I shall be unable to compete."

"It will be a crying shame if red tape stops Kingsford Smith from flying," thundered one newspaper.

The position was becoming desperate. The Australian authorities did their best when at long last the necessary data arrived from America. On September 28, the Civil Aviation Department carried out tests of the plane and found it entirely satisfactory. They granted Kingsford Smith certificate of airworthiness which would allow him to fly to England but might not be sufficient to let him take part in the race. "It will be necessary for the airman to satisfy the air race committee in London that the special certificate complies with the conditions of the race," they said.

This, naturally, did not satisfy Kingsford Smith. He cabled the Royal Aero Club in London asking whether the certificate complied with conditions. If the reply were favourable he would leave for England on September 29. If the reply were delayed, he would be unable to compete as loss of more time would make it impossible for him to reach England in time for the start. He then appealed direct to the Air Race Committee in Melbourne asking for authority to compete in the race. They replied they were unable to accept or reject certificates as they had delegated full power to the race committee of the Royal Aero Club in London. Sir Charles Kingsford Smith would have to await judgment from England.

The reply from the London Royal Aero Club came promptly. They accepted the certificate. Sir Charles Kingsford Smith could fly. Five hours later on September 29, only three weeks before the race was due to start, Kingsford Smith and P. G. Taylor set off in *Lady Southern Cross* for England. Disaster struck early. When they landed at Cloncurry, Queensland, they found a dent in the aluminium engine cowling and 14 splits, in some cases extending to the rivets. As Smithy explained later, if part of the cowling broke away, it could shear off the tail. There was no alternative but to limp back to Sydney for repairs.

Smith and Taylor were still determined to fly if repairs could be rushed in time. "I will have to get to England inside a week," said Smithy, "if I am to do any good in the race." The damage, however, was greater than they thought. There was no alternative. All Australia was shocked when it was announced on October 4, that Sir Charles Kingsford Smith had reluctantly withdrawn from the Centenary Air Race. Wing Commander Wackett, who was supervising repairs to *Lady Southern Cross*, said that he could not possibly get the plane ready to start before October 6. That did not give Sir Charles a chance to start in the race with the engine tuned up

in England. It would be far too risky to attempt the flight
without this being done.

Despite this justification, a howl of rage rose from critics
against this man who had risked his life in the air scores of
times and had won many records and much fame for
Australia. Some said he never intended to pit himself against
these world champions. He had lost confidence in himself
and his machine, or both. He would fly only if he was sure of
success. He was no sportsman. Some accused Smithy of
' squibbing". "A nation's hero may often become a nation's
whipping boy overnight," wrote Smithy later.

Kingsford Smith pondered his problem. He decided the
best thing he could do was sell *Lady Southern Cross* in
America and repay his sponsors. The cheapest way was to fly
her there. He and Ulm had been the first to fly the Pacific
from east to west. Now he and Taylor would be the first to
fly from Australia to California. Experts said he was
ill-advised to attempt such a hazardous flight in a land plane.
There was some suggestion that the Government should stop
him.

Meanwhile Smithy replied vigorously to his critics in
Smith's Weekly, one of Sydney's most trenchant newspapers.
"I've done some foolish things in connection with the big air
race," he wrote. "I admit them. But I'm no squib. I know
there are people who say I am. I know there are others who
contend I am pleased to be out of the race; and others, again,
say I'm wholly and solely to blame for being out. In short, I
know I am in the gun with a lot of good people who
previously thought well of me." He went on to explain that
he was flying the Pacific in a bid to repay his sponsors and
told his "squib critics" to have a look at the map and see
whether an England-Australia flight looks any worse than a
flight across the Pacific.

"Well, I'm putting my cards on the table — I'm saying my
piece," he continued. "I'm out of the race. That's a punch in
the solar plexus. But worse is that squib talk. That's hitting
below the belt. Anyway I'd like anyone who says I'm a squib
to say so in my hearing. And don't get the idea I'm thinking
of legal action." Smithy then explained how he tried to get a
British Comet and declared he would not have had a dog's

chance against the others without a variable-pitch propeller. For that reason he had bought an American plane.

"And while I'm on the subject of an American product," he continued, "I'd like to hazard the view that a lot of people who criticised me for choosing an American job are driving American cars. Yes, I'll admit I'm bitter on this subject. There's no howl of protest in England because Malcolm Campbell races a French car and no less a person than Earl Howe, President of the British Racing Driving Club, races Italian and French cars.

"Getting back to the race itself," he continued, "Detroyat, aerobatics champion of Europe and a French idol, entered an American plane, and K.L.M., Holland's leading airline company which is strongly supported by the Dutch Government, are banking on an American job. But because, like these other people, I chose a plane which I think will give me the best possible chance, I'm unpatriotic. Bunkum!"

Smithy admitted frankly that Captain Johnston had advised him to get the necessary certificates in America and said he was misled when Lockheed's test pilot flew the plane to the Department of Commerce airport near Los Angeles and told him on return that the machine was OK.

Referring to his return from Cloncurry, Smith wrote: "I know there are critics who assert that a small matter like cracks in the cowling would never have held me up if I didn't want to be held up. I know they've been saying that cracks in the bonnet would not hold up a car. Maybe not. If a bit of bonnet comes adrift it would not shear away a rear wheel. But if a bit of cowling breaks off — and it certainly threatened to do so on the Altair — it would be immediately whisked into the slipstream and bashed against the tail. If you've got any imagination, you may be able to figure out what a piece of metal travelling at anything up to 280 miles an hour could do to a vital part of a plane. I may be a mug, but there are limits to the risks I take."

"I had my cowling fixed," continued Kingsford Smith. "And I frankly admit this — I could have got to England in time. But after seven flights along the route, I know what an all-out trip such as was necessary would have meant to the human element as well as to the engine. For all our months of physical training, Bill Taylor and I would have been 'goosed'. And the engine and the plane, too, for that matter would have needed a complete overhaul, for which there would not have been time. If there had been a new engine waiting for me, that would have been a different matter. But there was no reserve engine — price put it out of the question. And the flight to England and then the race out again would have cost over £700 for fuel alone.

"And supposing we did get through the race," continued Smith, "which, in view of what our physical condition would have been after the sprint home, was very, very doubtful, what then? I could not sell the 'bus here because it cannot get a commercial ticket. And what return can I show my backers? It is primarily because of my backers that I am tackling the Pacific flight. They are going to be paid. If I pull this job off, there will be money for them and for me; and I'll certainly be able to sell the Altair in America. Anyway I've put my cards on the table. I did my best but the fates were against me. I'm sorry."

This reply to his critics brought a mass of mail to Smithy. Some of it was couched in language that would have shamed a bullock-driver. One envelope contained a white feather. Other letters, however, condoled with him on his ill luck and expressed the writers' enduring faith in his courage and integrity. *Smith's Weekly* summed up the feeling of his host of friends when it addressed him in Candid Communication:

"Dear Kingsford Smith,

One thought on reading your open address to the people of the Commonwealth! You say that you are making this Pacific flight to rehabilitate yourself with your fellow Australians. Such a flight carries with it tremendous risks alongside which the England-Australia route is a mere nothing. Need you take them? Australians do not demand such a gesture from you. They know you as the greatest airman the world has produced — from what you have already done,

Yours,
Smith's Weekly."

This and other appeals had no effect on Smithy. He was

The first woman to fly from England to Australia, Mrs. Keith Miller, with her co-pilot, Captain W.N. Lancaster. Their flight was also the first in a light aeroplane. Captain Lancaster later perished of thirst in the Sahara desert while trying to break Amy Johnson's record for a flight from England to the Cape.

Beinhorn, the second woman to solo from Europe to Australia (1932).

Expert mechanic Tommy Pethybridge, who vanished with Sir Charles Kingsford Smith in the Bay of Bengal while on a flight from England to Australia in 1935. Wreckage from their plane was found two years later.

C.W.A. Scott, the Qantas pilot, who, with T. Campbell-Black, won the Melbourne Cent Air Race.

G.U. ("Scotty") Allan in R.A.A.F. Uniform. One of the great pilots of early Australian Aviation. He became General Manager of Qantas.

C.T.P. Ulm, Kingsford Smith's co-pilot on m his historic flights.

determined to rehabilitate himself in the affection of all people and to repay his backers.

Thus it came about that on October 20, 1934, the stage was set for two of the world's greatest aviation events. Sixty thousand people crowded Mildenhall (England) airport to see 20 aeroplanes, representing six nations, take off in the great Centenary Air Race to Melbourne, Australia. Half a world away, at Brisbane, a much smaller crowd saw Kingsford Smith and P. G. Taylor set out next day in the *Lady Southern Cross* on the first west-east crossing of the Pacific.

The number of contestants in the great race had dwindled considerably since July when it was announced that there were 64 entries. Wiley Post had ruined the engine of his plane *Winnie Mae* in one of the earliest stratosphere flights. The French ace, Detroyat, was out. Louis Bleriot withdrew his plane with the retractable undercarriage. Harold Gatty and Harry Lyon also dropped out. The all-Australian plane was not ready in time. The promised American fleet of planes had dwindled to three with Roscoe Turner and Clyde Pangborn sharing instead of each flying his own. Of the 20 left, the three Comets were the favourites to be flown by Scott and Campbell-Black (the most fancied), the Mollisons and by Cathcart Jones and Ken Waller, whose Comet was entered by the Australian, Bernard Rubin.

Jim Mollison and his wife, the redoubtable Amy Johnson, were first away from Mildenhall in their black-and-gold Comet. They shattered all records to Bagdad and India, burned out an engine and abandoned the race. "We have put all our money in this machine," said Amy who was at the controls when a piston and gasket broke. She admitted she shed a few tears.

It was soon apparent that Scott and Campbell-Black would win if their tortured engines held out. They were ready to take grave risks. Ignoring the official warning to cross Northern India and follow the Malaya coast, they headed hell-for-leather from Allahabad to Singapore, 1400 miles of it over storm-tossed water. Scott's philosophy was summed up in the revolver he carried in case they came down in the Timor Sea. "The feel of my revolver is my abiding comfort," he said. "If I came down in the Timor, well the sharks would soon get me, and you couldn't call that suicide, could you?"

Plodding staidly behind the leaders came the Dutch air-liner piloted by K. D. Parmentier and J. J. Moll who explained blandly that they were not really racing. They just wanted to prove the possibility of long, fast flights by ordinary commercial aircraft. They carried four passengers, flew the normal Dutch Amsterdam-Batavia air route and delivered mail on the way.

Meanwhile the trail behind was dotted with wrecked and disabled planes. Australian James Wood and Flying Officer D. C. Bennett were unlucky when they brought their Lockheed Vega down on soft sand at Aleppo. The under-carriage collapsed, the plane overturned and they were out of the race. American Jacqueline Cochran and pilot W. Smith were forced down at Budapest by wing-flap trouble and damaged their plane. Australian Ray Parer and G. E. Hemsworth, flying a Fairey Fox, were forced down in France with radiator trouble, abandoned the race, but continued in leisurely style to Darwin where they arrived in February. Americans J. Wright and J. Polando, in a Lambert Monocoupe, withdrew at Calcutta, while Flight Lieutenant Shaw's Klemm Eagle was damaged at Bushire. Brake trouble forced Captain Neville Stack and S. L. Turner, flying an Airspeed Viceroy, to withdraw at Athens.

Death lay in wait for New Zealand Flying Officer D. H. Gilman and his co-pilot, J. K. C. Baines, in the Italian Appenines while flying from Rome to Athens in their out-dated Fairey Fox. They apparently had engine trouble while flying about 90 feet above the mountain tops. They plunged 5000 feet to crash 100 yards from a landing ground which they were trying to reach. Their plane caught fire and their bodies were burnt. Further along the route Asjes, Geysendorfer and Pronk had a narrow escape from death when the retractable undercarriage of their Pander S4 failed to reopen at Allahabad. They made a belly landing, completely smashing the faulty undercarriage and damaging a wing and a propeller. They rushed repairs and were taking off again when their plane collided with a motor-car carrying a beacon light. The undercarriage was torn off, a petrol tank burst and the plane was soon a mass of flaming wreckage. The three men scrambled clear with slight injuries.

Meanwhile, more than 700 at Archerfield, Brisbane, saw Kingsford Smith and P. G. Taylor take off for the perilous flight across the Pacific. The airmen arrived at the airport to the strains of "For they are Jolly Good Fellows". Before climbing into the plane, Kingsford Smith spoke into the microphone. "Cheerio, Australia," he said. "I'll be back as soon as I can." The plane took off and vanished quickly in the darkness. They touched down to a grand reception at Suva, Fiji, having covered the 2000 miles over the ocean in 12 hours. News of their feat, however, was swamped by reports of the great race.

These reports roused Australian enthusiasm to fever heat as Scott and Campbell-Black, in their red Comet, limped the last 100 miles over the Timor on one engine, made quick repairs at Darwin and swept into Melbourne only two days, 23 hours after leaving England. A crowd of 50,000 threw up their hats and sent up a roar of cheers as they flashed across the finishing line at Flemington racecourse before touching down at Laverton. The two airmen were cramped and almost reeling from fatigue. They had gone without sleep since leaving Mildenhall and had three days' growth of beard. Exhaustion showed in heavy rings round their eyes. One of the first to grasp their hands was the New Zealand flier Jean Batten. They thankfully accepted two bottles of beer and two packets of sandwiches which Brisbane airwomen Mrs. H. Bonney and Miss Peggy Doyle had thoughtfully brought for them. They were somewhat refreshed, therefore, when they flew in two Moth aeroplanes to receive an ovation to the strains of "See the Conquering Heroes Come" from the crowd at Flemington.

Second, ten hours after the victors, came the sedate Dutch airliner, helped by the people of Albury who dug it out when it bogged in an emergency night landing there. The Dutch gave up the second prize to win the handicap section. Queen Wilhelmina was so delighted with their success that she knighted members of the crew.

Third home, to rank second in the speed section, were the American aces Roscoe Turner and Clyde Pangborn in their

Boeing, followed in succeeding days by Cathcart Jones and
Ken Waller, plagued by engine trouble, New Zealanders
MacGregor and Walker in a Miles Hawk, the Stodarts, flying
an Airspeed Courier, and young South Australian Jimmy
Melrose, who glided his tiny Puss Moth a mile over the Timor
Sea into Darwin with his petrol tanks empty to come second
in the handicap section. After Melrose came the Danes,
Michael Hanson and D. Jensen, in their Desoutter, New
Zealanders J. D. Hewett and C. E. Kay in a DH Dragon
Rapide, H. L. Brookes and Miss Lay in a Miles Falcon and
Davies and Hill in a Fairey 3F.

All Melbourne turned out for the victory parade given on
October 31, when only seven of the finishing planes had
flown in. The airmen rode in motor cars through storms of
cheers and confetti. At times the enthusiasm reached almost
hysterical proportions, especially when Jimmy Melrose, the
baby of the race, who had made a special rush from
Charleville to be there, came into view. Melrose was quickly
recognised by his fair hair and boyish appearance. Women
struggled with police to shake him by the hand while some
tried to kiss him. The cars carrying the airmen were filled
with flowers.

While the Melbourne festivities continued, Kingsford
Smith and P. G. Taylor made the dangerous 3150-mile flight
over sea from Fiji to Honolulu. They brushed with death in a
blinding storm 15,000 feet up when Smithy accidentally
knocked the flap switch when turning on the landing light to
check up on the rain. *Lady Southern Cross* went into a wild
spin, hurtling seawards through the storm, her klaxon
screaming. Both airmen took it in turns struggling with the
controls. The plane plunged 9000 feet before Smith regained
control, realised the error and rectified it. "We thanked God
for so much altitude," he said after American army pursuit
planes had escorted them into Wheeler Field, Honolulu.
"Otherwise we would have met our doom." Honolulu draped
them with leis, gave them a motor-cade ovation and
entertained them royally.

It was here that Kingsford Smith paid the first of many
tributes to his navigator, P. G. Taylor. "Taylor is a wizard,"
he told admirers. "With Harold Gatty, he is the best in the

world. He found the exact point aimed for in the Hawaiian group." Later he was to say that he would fly anywhere with Taylor.

Fifty thousand people welcomed Smithy and Taylor at Los Angeles after they had landed at Oakland, California, on November 4, 1934, having made a far more hazardous flight than the England-Australia racing men who had followed a well-blazed trail. The general jubilations were halted briefly for Smithy when a Los Angeles promoter attached the *Lady Southern Cross* and filed an action against him for 2750 dollars which, he said, Smithy owed him in connection with the first Pacific flight in 1928. Smithy was most indignant. He called the action absurd, ridiculous and preposterous and threatened to bring a counter-action for damage, defamation of character and delays caused to his plane. The action was settled out of court for what Smithy's lawyers described as a nominal sum.

Smithy's plans still went awry. He abandoned his intention to fly across America and the Atlantic and then on to Australia, thereby again circling the world by the longest route. "There may be glory but there is little cash for aviation feats," he remarked sadly as he set out to try to sell the *Lady Southern Cross* in an attempt to balance his accounts. "We chaps who blaze air trails have little to show for our deeds when it comes right down to it. Ill fate has marked us for its own. But tomorrow is another day and I am not downhearted."

His words were prophetic. Ill fate marked many airmen for its own. Tom Campbell-Black was struck by the propeller of a taxiing aircraft and cut to pieces in 1936 a few months after he had married the actress Florence Desmond. Scott also died tragically. The sands were running out, too, for Smith and Ulm. Death lay in wait for them on the heroic scale.

Chapter 20

Death in the Pacific

In December 1934, Charles Thomas Philippe Ulm, who with Kingsford Smith had been the first to fly the Pacific, stood poised at Oakland, California, for another death-defying flight to Australia, more than 7000 miles away. In his brilliant, organising mind, Ulm still dreamt of owning a company that one day would send planes on regular schedules across the vast ocean.

For some years now Ulm had been battling. When the original A.N.A. company he and Smith owned collapsed, he bought the plane *Southern Moon* from the debris, had it rebuilt and planned to compete again for the Australia-Singapore leg of the Empire air route which had not yet been allocated. To strengthen his bid he set out in the *Southern Moon,* renamed *Faith in Australia,* to fly round the world and break the record, both for speed and for distance, set by the American Wiley Post, the first man to circle the globe in continuous planned flight. Post had completed the 15,500-mile northern hemisphere route in eight days, sixteen hours in 1931, with a Tasmanian, Harold Gatty, as his navigator. Ulm meant to fly by the longer empire route, crossing the equator twice.

In 1933, he took off with P. G. Taylor and G. U. (Scotty) Allan for England, the first leg. Bad luck beset them from the start. The petrol feed system failed while they were flying across Australia to Derby. They took it in turns pumping petrol to the motors. Ulm collapsed from petrol fumes and Taylor worked to revive him while Allan flew the plane. They picked up new pumps at Derby and sped on to northern India where a piston broke. They repaired this and, flying on, had to dump 400 gallons of fuel and land with smoking engines in Persia. They were forced to land again at Lyons. The journey to England took 17 days.

Stubbornly intent on the world flight, Ulm, Taylor and

Allan flew *Faith in Australia* to Portmarnock Sands, Ireland, for the still-perilous "hop" across the Atlantic. The 6000 lb. weight of petrol she needed was too great for the sand on which she stood. One wheel sank deep and the starboard wing dug into the beach. The undercarriage collapsed. *Faith in Australia* was bogged. Before they could dig her free, the tide came in and lapped around her. That would have been the end had not Lord Wakefield, the oil magnate, telegraphed Ulm to have the plane salvaged and repaired and send the bill to him.

Three months later *Faith in Australia* was ready again. It was too late to resume the world trip for they could no longer rely on favourable winds. Ulm, Taylor and Scottý Allan therefore flew back to Australia, cutting the record then held by Kingsford Smith to 6 days, 17 hours, 56 minutes.

Ulm did not win the Australia-Singapore contract. As has already been noted in a previous chapter, it went to Qantas. He was not dismayed however. He made several pioneer flights carrying the first official mail between Australia and New Zealand, and Australia and New Guinea, then set his sights on promoting the first regular air mail and passenger service across the Pacific between Australia and America. He planned to do this in short stages. Already, it was later revealed, he had built at his own cost, with native labour, a landing ground made of crushed coral on Fanning Island, a thousand miles south of Hawaii. If the great flight he planned proved this landing ground efficient, his company would consider construction of similar aerodromes on other islands in a great chain across the Pacific, linking Australia and New Zealand with America. It was vision on the grand scale, an enterprise that deserved success.

To prove his plan feasible, Ulm organised an all-British experimental survey flight across the Pacific, taking in his coral aerodrome on Fanning Island on the way. For the flight, Ulm bought a British twin-engined, low-wing Airspeed Envoy monoplane, christened it *Stella Australis* tested it at Portsmouth, and took it by sea to Canada. He proved it further with flights from Montreal to Detroit and on to Texas, Los Angeles and Oakland, California, where Smithy and he had begun their first great Pacific flight in 1928. Ulm, then only 37, chose as crew G. M. Littlejohn, chief instructor at the New South Wales Aero Club (co-pilot) and Leon Skilling a maritime navigator and radio operator, formerly in the service of the Orient Line.

Ulm was confident of success. He announced at Oakland that full financial and technical plans for a trans-Pacific service had been completed. "I expect Great Pacific Airways Ltd., of which I am managing-director, to establish a service in the next two years," he said. "Planes, one a week each way between Sydney and Honolulu to connect with the steamer service between the United States and Honolulu, would reduce the transportation time of 21 days from San Francisco to Sydney to 7½ days." The proposed service would not at first extend beyond Honolulu, he told another newsman, because there was no aeroplane at present capable of operating economically on non-stop flights between Honolulu and America.

On November 30, Ulm took *Stella Australis* up for a test flight loaded with 600 gallons of petrol. He returned to announce that she was in excellent shape. "We can make the hop with no trouble at all," he said.

Navigator Skilling also voiced his philosophy to a friend. "Familiarity with fear breeds contempt," he was reported to have said. "I believe there is a God and that we continue to live in some mysterious form after death separates us from our loved ones. Our lives are mapped out for us by God when we are born and no matter what we do we cannot change the results. If it is our time to die, we will die, but if the time has not arrived, we will live. The chance we are taking in flying across the Pacific is no greater than walking across a busy street or sleeping in bed. In one instance you may be killed by a motor car, while in the other you may die if an earthquake wrecks your house. You take these chances every day. We are familiar with the perils of the journey but, with the aid of science, our chances of meeting with disaster are no greater than with you who stay at home."

Ulm, Littlejohn and Skilling took off from Oakland in *Stella Australis* on the morning of December 4, 1934, bound for Honolulu, 2400 miles across the Pacific. They were confident that, if they ran into trouble, the Honolulu radio beacon would guide them safely in. Approaching Hawaii, they ran into cloud and had to fly by dead reckoning. All went well till they were reported to be within 480 miles of Honolulu. Then they lost their way. Skilling could be heard tapping out requests for a bearing and for the radio beacon to be turned on. The beacon was on, but they were not getting it. Somewhere over the vast Pacific, the plane was droning into the unknown asking for the precious beacon it could not get. Then came the fateful signal "We are landing in the sea. Please come and get us," followed by stark silence.

The story of the tragic flight is best told by wireless signals received from the flying plane. At first they were cheerful and confident. A radio amateur picked up a signal at 10.15 a.m. in which Ulm said they were making good progress; the weather was perfect and the engine was running sweetly. At noon he reported they had picked up a nice little tail wind. Two hours later they were passing over the steamer *Lurline*. "The engine is fine, the weather perfect. Starting lunch. We do not expect to get our feet wet," reported Ulm. By 4.30, Ulm reported that the plane was nearly half way to Honolulu. At 12.25 a.m. on December 5 the liner *President Coolidge* radioed San Francisco that Ulm was 500 miles from Honolulu. At 6.30 the American Army Air Service reported that Ulm was asking for weather reports for Wheeler and Luke Army Airports, and that he expected to arrive about 8 a.m.

Then the signals went haywire. The first indication that the airmen were in trouble was a message picked up at 7.30 a.m. saying they were lost and running short of petrol. They sent repeated calls for the radio beacon to be turned on. "He keeps asking for the beacon," wailed the operator. "The beacon has been on since midnight." Experts revealed later that the radio beacon sent signals on a directional beam and, if the airmen were badly off course, they would not get it.

Another message now revealed that Ulm did not know in which direction he was flying. "I don't know if I am south or north of the islands," he said. "Must get beacon soon. Have no position. Must be badly lost." Soon he was reporting that he had only 45 minutes of petrol left. "Trying to pick up

land," he said. The American navy sent up planes to circle
Oahu and keep watch for him. To cause more confusion
came the message. "We are south of Honolulu but are
heading back towards our plotted course." Then came the
SOS calls with the final message at 9.23: "Going down into
sea. Plane will float. Come and get us."

The request was not an easy one. If the Honolulu
authorities knew where they were, there would be strong
hopes of rescue. Ulm had frequently said that, if he were
forced down, he and the crew would climb on to the wooden
wing and cut the engines free. Buoyed up by empty fuel
tanks the plane, he maintained, would float indefinitely
providing seas were calm. Most conceded they had at least
two or three days to find them, but where were they? Ulm in
one of his messages, had said he was south of Hawaii and
heading back towards his plotted course; an earlier message
indicated he did not know where he was. American Navy
officers thought he had come down short of Oahu, while the
Navy meteorologist, basing his prediction on the plane's
speed and the prevailing winds, was of opinion that *Stella
Australis* could be floating anywhere up to 300 miles
north-east of the island. Others suggested that the airmen had
overshot Hawaii flying blind through the overcast sky. To
complicate matters further, radio men now believed that the
growing weakness of the signals was not due to failure of the
radio equipment but to the distance off course of the lost
airmen, all of which added up to the fact that somewhere
within a radius of 300 miles in that mighty ocean around
Honolulu, three gallant Australian airmen might still be afloat
on a flimsy plane that was not even a pin-point in a waste of
water.

Faced with a welter of conflicting views, the American
army and navy forces based on Hawaii mounted a massive
search for the missing airmen on the grand scale the
Americans use when they have an urgent job to do. As many
as 18 submarines, three destroyers, several minelayers and
coastguard cutters, searched islands and seas where it was
thought the plane might be drifting, while squadrons of army
and navy planes criss-crossed 250,000 square miles of ocean.
The whole Japanese fishing fleet based on Hawaii kept a

sharp look-out, spurred by the offer of the Australian Government of 5000 dollars for the discovery of all or any of the aviators alive and 1000 dollars for the finding of a body or wreckage of the plane.

Meanwhile, in Australia's Federal Parliament, Prime Minister J. A. Lyons revealed that the Commonwealth Government had guaranteed the expense of Ulm on his trans-Pacific flight to the extent of about £8,000. The Government did this, he said, because the airman was anxious to demonstrate the success of a British plane. Ulm had everything planned carefully so that success seemed reasonably sure. The Government was influenced by Ulm's ability and experience in organising long-distance flights.

Later, in the Senate, Cabinet Minister Sir George Pearce said the Commonwealth's guarantee of £8,000 for Mr. Ulm's flight was made in an endeavour to speed up the establishment of a commercial air service across the Pacific. Mr. Ulm had rendered great service to Australia and was one of their most reliable pilots. His flight differed from that of Sir Charles Kingsford Smith in that it was planned definitely to pioneer a trans-Pacific service. The Government later made .a grant of £5000 to Mrs. Ulm.

As the days passed hopes of finding the airmen faded. The naval and air forces were recalled to base. A private search sponsored by Mrs. Ulm and the State Government of islands north-west of Hawaii failed. One of Australia's greatest airmen and his crew had been swallowed by the sea.

— Chapter 21 —————————————

The tragedy of Kingsford Smith

Following the loss of C. T. P. Ulm and his men, the spotlight turned increasingly on Ulm's old friend and comrade Charles Kingsford Smith, who was still in the race to establish an air-line to New Zealand and across the Pacific. To strengthen his claim Kingsford Smith agreed in 1935 to carry special mail to New Zealand as part of the Jubilee celebrations of King George V and Queen Mary. With P. G. Taylor, relief pilot and navigator, and John Stannage, radio operator, Smith took off for New Zealand in what he intended to be the last overseas flight of the old *Southern Cross*. First they ran into bad weather, then, six hundred miles out, one of the exhausts broke. Flying metal smashed the propeller of the starboard engine which threatened to shake the plane to pieces.

On two engines, Smith turned back. To lighten the plane and maintain height he threw overboard luggage, tools, freight and dumped a large quantity of petrol. Smith had nursed the crippled *Southern Cross* to within 200 miles of Australia when peril struck again. The tortured port engine burned up its oil. The pressure gauge fell alarmingly. It was only a matter of minutes before the engine would burn itself out and, with only one engine, *Southern Cross* would plunge into the sea.

It was then that P. G. Taylor performed the feat that won him the Empire Gallantry Medal, later changed to the George Cross. The only way of saving the plane was to get oil to the port engine. In the useless starboard motor were nine gallons of perfectly good oil. Taylor set out to transfer the oil from the starboard motor to the port. To do this he took off his boots, climbed out of the cabin, pressed his toes to a streamlined strut and, with shoulder thrust firmly against the wing, edged his way to the starboard engine. The gale from the airstream tugged and threatened to hurl him into the sea, raging only a few feet below. Clinging to the engine cowling, Taylor drained oil from the useless engine into the metal casing of a thermos flask Stannage passed to him. He then made the perilous trip back to the cabin.

The worst part of the job was still to be done. This was to get the oil to the port engine, not far now from burning out. Taylor could not walk the strut in the blast from two engines, so Smithy cut off the port engine. The *Southern Cross* slid steadily towards the sea as Taylor edged himself along the strut. He was clinging to the cowling when Smithy opened up full throttle again and lifted the *Southern Cross* away from the sea only a few feet below. Taylor then poured the oil into the tank of the port engine. John Stannage gave him the thumbs up sign — the oil pressure had lifted and they were safe for a while. Taylor walked the strut back to the cabin.

Meanwhile, their friends at Mascot had been following the course of the flight with increasing alarm. In a grim running commentary, Stannage had reported the accident to the propeller and the burning up of the oil. The pilot ship *Captain Cook* left Sydney to patrol along the track of the crippled plane. H.M.S. *Sussex* prepared to put to sea. Meanwhile, Smithy, Taylor and Stannage battled to keep the *Southern Cross* in the air. They were forced at last to jettison the mail. The port motor ate up oil. Every half-hour Taylor made the perilous trip along the struts taking oil to the labouring engine. By this means, coupled with Smithy's brilliant flying, he kept *Southern Cross* in the air till at last she touched down safely at Mascot.

Kingsford Smith ran into a welter of ill-luck. He could not sell the *Lady Southern Cross* in which he and Taylor had flown the Pacific to California. Smithy had the plane shipped to England where the authorities refused to give him a British ticket of airworthiness which would have increased the plane's value. His plan to get British finance for an air-line to New Zealand, linking with an American line at Honolulu, failed. The Australian Government quibbled about paying an advance on the old *Southern Cross* which they had agreed to buy. A request to Australia for funds was refused.

Kingsford Smith was worried. All his plans had gone

astray. He believed officialdom generally was against him. He had just recovered from a severe bout of influenza and was not well. Under this weight of anxiety, Sir Charles Kingsford Smith decided to get home quickly and the best way was to fly, breaking Scott and Campbell-Black's London-Melbourne record if he could. His wife appealed to him from Australia not to make the flight. Friends said it would be rash to attempt it in a single-engined plane.

Smithy, however, was adamant. He just had to get *Lady Southern Cross* back to Australia. He compromised by promising that this would be his last major overseas flight. "All I want now is an administrative job at a desk with a nice glass top," he told British newsmen when they asked if he would take an active part when a trans-Tasman mail service began. Even then he ran into strife. The British authorities, strictly applying the international safety regulations, would allow him to load only 138 gallons of petrol instead of the maximum capacity of 400 gallons. This would take him only as far as Marseilles and would increase the number of stops if similar restrictions were imposed along the route.

Accompanied by 28-year-old Tommy Pethybridge, brilliant aero mechanic and chief of technical training at the Kingsford Smith Flying School, Smithy took off from London on October 23, 1935, for Melbourne. A large crowd including C. W. A. Scott, Jimmy Melrose and H. F. Broadbent, all of whom were about to start record-breaking flights to Australia, saw him off. He announced that he hoped to make a quick but not necessarily record flight. His main aim was to demonstrate that a fast air-mail service with a full payload was possible.

Kingsford Smith did not get far on that attempt. He ran into hail and snow over the mountains of Greece and could not see ten yards ahead. Ice, forming on his torn wings, threatened to send him crashing out of control. Flying blind, he turned and ran before the storm, making for the nearest aerodrome at Brindisi. With visibility nil, he was lost till at last he picked up the beam of a lighthouse. The 'drome at Brindisi had no illumination for night landing. He radioed his dire need to land and circled round till the Italians rounded up enough motor cars to illuminate a runway with their headlamps and thus enabled him to make a safe landing. When he returned to Croydon two days later, officials said they had not seen a machine in worse condition and still flying. The leading and trailing edges of the wings were torn and the woodwork badly affected by hail, snow and ice.

On November 6, 1935 Kingsford Smith and Pethybridge were ready again. This time Smithy made some slightly-conflicting announcements, possibly with an eye to the authorities still clamping their petrol restrictions on *Lady Southern Cross*. "I want to insist," said Smithy, "that the primary object of my flight is not to break the record but to employ the most logical and easiest means of delivering a machine to Australia. I intend to make the fastest possible trip because I believe that no flight is valuable unless it demonstrates that it is a more rapid means of transit than the existing forms. If I am fortunate enough to arrive in Melbourne in less than the air-race time, it will be purely accidental. I realise I would have to put in a month's preparation and incur considerable expense to stand a reasonable chance of bettering Scott and Black's achievement. I am able to do neither. Moreover, I will be unable to load sufficient petrol to make long hops."

Elsewhere he let it be known that, although he would leave England with only 138 gallons of petrol as decreed by the authorities, he would take on the maximum of 400 gallons at Marseilles or Athens. "If the early stages of the flight are satisfactory," announced one newspaper, "Sir Charles intends to try to reach Melbourne by noon on Saturday (November 9) which would beat Mr. C. W. A. Scott's record."

Thus, without full preparation, spurred possibly by a desire to prove to his former critics that he could have won the great air race in his American machine if fate had not stopped him, Charles Kingsford Smith set out with young Tommy Pethybridge to speed half across the world in a final bid for the England-Australia record. Just before the take-off from Lympne he was informed that it was snowing in northern Italy, while thunderstorms threatened elsewhere along the route. "There'll be no turning back this time," said Smithy. "I must stick up somehow."

They reached Allahabad 29 hours, 28 minutes later, less

than three hours behind Scott and Black's time and sent a message to Sydney that they hoped to be in Darwin by midnight on November 8 and leave 30 minutes later direct for Melbourne.

They took off from Allahabad and were never seen again.

The news of the disappearance of the man many hailed as the world's greatest aviator came as a shock to Australia, Britain, America and other air-minded nations. The only clue to the course Kingsford Smith had taken came from young Jimmy Melrose, then well advanced on his bid for the England-Australia solo record. Melrose was flying his Percival Gull over the Bay of Bengal in the darkness about 3 a.m. on November 8 when another plane, travelling at twice his speed, sped 200 feet above him. Melrose recorded an uncanny feeling as he saw the spurts of flame from twin exhausts over a desolate ocean where he thought himself alone. The other plane could only have been the *Lady Southern Cross*. Even with this clue the authorities could do no · more than announce that the airmen were missing between Akyab in Burma and Victoria Point, 750 miles south on the Malay Peninsula.

Within hours of the announcement that Kingsford Smith was overdue, Royal Air Force flying boats and bombers based on Penang, Taiping and Singapore were searching for the missing airmen. Wireless messages were sent to ships at sea to keep a sharp look-out. Jimmy Melrose abandoned a good chance of winning his record bid to look for his hero, Kingsford Smith, while Qantas ordered crack pilot Scotty Allan, friend and companion of Kingsford Smith in many adventures, to join in the search in their emergency air-liner then based on Singapore.

· It was like looking for a needle in a haystack. No one could even guess where the missing men might be. Monsoonal storms could have forced them down in the sea. They could have made a forced landing on any beach down the long stretch of coast or be marooned on one of the maze of islands off the Malay Peninsula. They could just as likely have crossed the coast and crashed or force-landed in impenetrable jungle anywhere in Burma or Siam.

Many, however, considered there was a good chance of

"Faith in Australia", C.T.P. Ulm's most famous plane, which pioneered many air-routes. This photograph was taken at the dispersal bay of a northern aerodrome.

P.G. (now Sir Gordon) Taylor, with Kingsford Smith and radio operator John Stannage. The three had a nightmare flight in 1935 when they set out from Mascot to take the Jubilee Mail to New Zealand. A smashed propeller forced them to turn back. P.G. Taylor kept the aircraft in the air and saved their lives by climbing out of the cabin and walking struts in the airstream to get oil from the useless starboard motor to the port motor which threatened to burn out. For this he was awarded the George Cross.

finding them alive. Lady Kingsford Smith, staying in Melbourne with her mother, though distressed by the absence of reports from her husband, said she was confident he was safe. "He has been in many difficult situations," she said, "and his ability has always pulled him through. I have the utmost confidence in him and I am sure that, if he has met trouble, he has made a safe landing."

Meanwhile, as the days passed, fears grew for the aviators despite an assuring message from the Viceroy of India, Lord Willingdon, that, if they had alighted on one of the islands of the Mergui Archipelago, it might be several days before they were discovered.

One airman who had just flown from England, told of violent storms which nearly forced him down over Burma. He thought that, if an aircraft travelling at 200 m.p.h. were driven through one of those storms, either the engine would be torn from the fuselage or the wings would collapse. It would be like flying a machine into solid water.

The Dutch flier, Ever Vandyck, who had returned to The Hague from the East Indies, said he warned Kingsford Smith in Athens of gales that had raged for more than a week in the Akyab area. Kingsford Smith would cross the mountains at night and the slightest engine trouble would place him in jeopardy. Vandyck's theory was that Kingsford Smith's plane had been struck by lightning.

Those who thought the airmen might have crashed or been forced down in the dense tropical jungle of Burma or Siam doubted if they would be able to find their way through it to safety. They conceded there was hope, however, when a newspaper reminded its readers of the case of James Matthews and Eric Hook who force-landed in dense Burmese jungle while trying to fly to Australia in 1930. After a week's intensive search for them, by air and sea, hope was abandoned. This was premature for their plane in landing had become lodged in a clump of bamboo which held it for a while before letting it slide gently to the ground. Neither Matthews nor Hook was injured. They set out to follow a river looking for a village. Hook, suffering from fever, collapsed and could not go on. Matthews pressed ahead seeking help and himself was close to death when he came to a village. A rescue party of villagers found Hook dead. Matthews was taken by boat to Prome where he was hailed as a man back from the dead.

"The fact that Matthews and Hook survived their forced landing," continued the optimistic newspaper, "holds out some hope that Sir Charles Kingsford Smith and Mr. Pethybridge may have made a safe landing and may be sighted by searchers from the air. If they have done so, they may even now be in the hands of friendly villagers as Matthews was, and in that case it might be days yet before news would reach any of the larger settlements."

On November 16 came word that a plane had been seen flying low 10 miles inland on the west coast of Siam. A squadron of bombers and two flying boats flew over the area, while a land party searched the jungle. It became evident later that this plane was flown by H. F. Broadbent, then making his record-breaking flight to Australia.

A native on the border of Siam and Malaya then said that he saw a plane flying towards the mountains, where, according to another native, it had "shot up over a hill and dropped in flames." The search turned to this area without result. A ship reported that it had seen flares on Sayer Island, near Victoria Point. Aerial reconaissance found only native fishing boats.

Meanwhile Scotty Allan in the Qantas air-liner had searched a large area of jungle and had dropped pamphlets in various native dialects on villages, offering money for news of the missing airmen. On November 24, it was unofficially stated that the search had already cost more than £25,000.

As the days passed, hope faded. The searching planes were withdrawn. Jimmy Melrose flew on to Australia. The conviction came slowly that Kingsford Smith and Jimmy Pethybridge had met death in ocean or jungle. For two years their fate remained a mystery. Part of an undercarriage was then found washed ashore on an island off the Burma coast. Some of the tree tops on the island were shorn of branches. Experts believe the *Lady Southern Cross*, flying low, struck the top of the trees and plunged into the sea, taking with her Australia's greatest airman and his young companion.

— Chapter 22 ————————————————

War again

As the months passed, the clouds of approaching war began to thicken over the world. Though war would obviously flare first in Europe, there were signs that it would spread quickly to the East where a militant Japan was riding rough-shod over China, casting covetous eyes on the riches of Malaya and the East Indies and showing an increasing disregard for the other great powers, including Britain and the United States. War spreading over the East would cut the aerial lifeline between Australia, India and Britain. There was pressing need for another route and the only alternative was via the islands dotted about the Indian Ocean between Australia, India and South Africa.

One of the first to see the need for this was P. G. Taylor, pioneer pilot and navigator, hero with Kingsford Smith of the first west-east crossing of the Pacific and, with Ulm and Smith, of record-breaking flights between Britain and Australia, and Australia and New Zealand. Through the Federal Treasurer, R. G. (now Lord) Casey, Taylor convinced the Australian Government that a chain of bases across the Indian Ocean was imperative for defence. With the backing of the British and Australian Governments, he set out in June 1939, to survey possible island bases and in the process made the first crossing of the Indian Ocean from Australia to Africa.

For this survey, Taylor chartered from an American scientist, Richard Archbold, the P.B.Y. Catalina flying boat *Guba* in which Archbold and his party crossed the Pacific to New Guinea to collect specimens for the American Museum of Natural History. Archbold and his crew accompanied Taylor on the survey flight, together with Jack Percival, a Sydney journalist and aviation expert, who had flown with Kingsford Smith across the Tasman. They were unlucky on the first leg. They flew through torrential rain and lightning

with neither sight of sun nor stars, missed Cocos and turned
north-east to Batavia. Though plagued again by cloud, they
arrived at Cocos a few days later and, making the necessary
surface surveys, flew on to Diego Garcia in the Chagos
Islands, Mahe in the Seychelles, and Mombasa in Kenya
where the proposed new route would enter Africa.

Thanks to Taylor's foresight, the survey was completed
just in time. The chain of islands, together with the Maldives,
became bases for the Navy and Royal Air Force in their war
on raiders and submarines in the Indian Ocean. From them,
too, flew the Catalina flying boats which found lifeboats
packed with survivors from torpedoed ships and guided
rescue vessels to them. In this way more than 1,000 men were
saved. Since the war, the island bases across the Indian Ocean
have been used for regular air services by Qantas and South
African Airways.

Meanwhile, from the earliest days of the war, Australians
fought the enemy in the air above Britain, France and the
High Seas. Some who were on exchange duty with the R.A.F.
when Hitler struck, flew straight into the conflict. Australia's
No. 10 Squadron, arriving to collect nine Sunderland aircraft
for the R.A.A.F., were thrown at once into the battle,
protecting convoys which were running the gauntlet of
U-boats to bring arms and food to Britain.

When the Nazis invaded Norway, Australians struck
further afield. Flying-Officer Hurley, of Randwick, and
Flying-Officer J. H. Horan, of Kensington, Sydney, bombed
oil tanks at Bergen, while Flt.-Lt. D. H. French, of Brighton,
Vic., was in a squadron that attacked the German cruiser
Konigsberg. Australians flew in almost suicidal low-level
attacks against German panzer columns when the enemy
broke the French and Belgian lines at Sedan. With Dunkirk
falling, No. 10 Squadron helped throw a protective
"umbrella" over the hundreds of craft that snatched the
British army from the beaches to fight again.

Australians also were prominent among the "Few" who
fought off the Luftwaffe hordes in the Battle of Britain and
saved the homeland from invasion. Four Australians —
Flying-Officers R. Glyde, of Perth, J. Cock, of Renmark, J.
Curchin, of Hawthorn, and Flt-Lt. P. Hughes, of Cooma —

destroyed more than 30 German aircraft in the battle. Cock was shot down with his plane on fire. He baled out, swam ashore and was quickly in the air again.

As the months passed and Britain, now alone against the enemy, began the blood-toil-and-tears defiance that slowly led to victory, more Australians arrived to fight for freedom in the air. The most famous was Australia's No. 452 Spitfire Squadron, a group of high spirited young dare-devils welded into a venomous fighting machine by Britain's No. 1 ace, Brendan (Paddy) Finucane. With such fighting men as Keith (Bluey) Truscott and Keith B. Chisholm, 452 was the hardest-hitting unit in the Empire air force, with, at one time, three times as many kills as its nearest rival.

Finucane taught them how to fight and live in the air, how to spot enemy planes before they themselves were spotted and then how to outwit them. He had apt pupils. "All Australians are mad," he said once. "Waggle your wings and they'll follow you anywhere." Red-headed Truscott was the star of 452 squadron. He became so famous that the Marquess of Donegal launched the Red-head Spitfire Fund in his honour. Red-heads the world over sent £5000 which was said to have paid for the Spitfire in which Bluey bagged six Nazi raiders.

A long odyssey lay ahead for Keith Chisholm. A Newington boy, Chisholm had downed six when enemy fighters shot the tail from his plane near Dunkirk in October 1941. A German launch pulled him from the sea. They sent him to Lamsdorf prison camp in Czech Sudetenland. From then Chisholm had one aim in life — to escape. He got away the first time by prising up floorboards and squeezing through a ventilator. He and his comrades were betrayed by a quisling at Brno. He then teamed up with Douglas (Tin-Legs) Bader to get to a nearby airstrip in a prison work party, steal a Messerschmitt and fly to Britain. Bader's tin legs gave them away.

For the next attempt Chisholm teamed with two Australian soldiers who knew all about locks. They cut through a ceiling and walked the rafters to the boiler room, where the soldiers picked the lock. Scaling the fence, they scattered in a wheatfield. With three companions Chisholm made his way by night to Cracow, then to Warsaw, where he joined the underground forces. One of his companions was murdered; another died in the hands of the Gestapo. He and his friends struck down a German who challenged them and threw him through the ice of the Vistula. Chisholm and a friend bluffed their way by train through Berlin to Brussels and then to Paris, where Chisholm worked with the underground. On Liberation Day, August 25, 1944, Keith Chisholm reported back for duty at the American H.Q. of General Omar Bradley.

Long before then, however, the war had spread to the East. The Japanese smashed the American fleet at Pearl Harbour, swept through the Malay States to Singapore and were soon island-hopping through the East Indies towards Australia. With the air route cut, Qantas came into the battle. Their planes flew arms and supplies to the small Allied forces struggling desperately against odds to delay the Japanese drive through the islands. On the homeward journey they snatched women and children from threatened areas and flew them to the comparative safety of Australia. On these evacuation flights, some of the flying boats had a nursery atmosphere, with children playing on the floor and nappies strung on lines in the cabins.

Qantas pilots ran this shuttle service, first to Singapore and beyond, then, as the enemy advanced, to the Dutch East Indies. Always they were in peril from Zero fighters.

On January 30, Captain A. A. Koch left Darwin in the Qantas flying boat *Corio* bound, mainly with service personnel, for Sourabaya where he would pick up women and children waiting to be evacuated on the return journey. He was approaching Koepang in Timor when he heard a peculiar rattling in the fuselage and saw the air filled with tracer bullets. Passengers behind him in the cabin slumped dead as bullets ripped into the hull. Two engines burst into flames. Koch dived to water level and swerved and zig-zagged to avoid the flying bullets. He skimmed so close to the sea that at times the wing floats touched the surface.

Koch made for the nearest beach 15 miles away. With two engines on fire, the flying boat lost speed and he was forced to alight. The planing surface of the hull, however, was so badly holed that the drag from the holes caused the flying boat to pitch on its nose which plunged under water. Koch was thrown through the broken screen. He came to the surface to find the aircraft floating on its wings, its back broken and cabin windows at about water level. Looking up, he saw seven Japanese Zero fighters circling overhead. Koch found five others swimming around the craft. They were joined quickly by a naval officer who, despite the bullet holes in his chest and a gash in the head, struggled into a lifebelt, punched a hole through a cabin window and fell from it into the water.

No other sign of life came from the cabin which caught fire. Burning petrol spread across the water. The survivors had to struggle well clear of the blazing aircraft, some of the injured supporting themselves on floating debris. The flying boat had come down five miles from land. It was soon apparent they could expect no immediate help from shore. Captain Koch, whose leg was broken at the knee, and Mr. S. Moore, a passenger, set out to swim the five miles and seek help for their comrades. Three hours later they struggled up the beach, where Moore dragged Koch, on the verge of collapse from his injuries, above the waterline. They found no natives to send to the rescue of their comrades, three of whom struggled ashore in the next hour. Two were swallowed by the sea. Of the 18 who set out for Sourabaya in *Corio,* only five survived. With much difficulty Moore got through to Koepang. The Dutch sent one of their Dorniers with medical men aboard to rescue the others. Captain Koch, having survived several raids on Koepang, was flown back to Darwin in time for the blitz the Japanese launched on that port.

A month after *Corio* was shot down, the Qantas flying boat *Circe,* piloted by Captain W. B. Purton, vanished without trace while flying with evacuees from Tjilatjap, Java, bound for the new Qantas base at Broome. *Circe* was expected to keep in radio contact with a sister craft which took off from Tjilatjap a few minutes earlier. Not one message was received. *Circe* is believed to have been shot down by Jap fighters.

From then some of the Qantas planes, known as bully-beef

bombers, were the work-horses supplying the Australian fighting men who drove the Japanese from New Guinea. One landed with supplies at Milne Bay just before the Japanese forces swarmed ashore to be driven back by Australians and receive the first check in their advance.

Kittyhawk fighters of the RAAF, manned among others by Bluey Truscott, back from Europe, had a big share in the Milne Bay victory. They drove off 40 enemy planes that tried to seize the improvised air-strip on the day of the sea invasion and shot up the enemy transports. Truscott himself flew at tree-top height to silence a 37mm. gun harassing Australian defenders.

Meanwhile, in Europe, Australian and British airmen were carrying the war into the heart of enemy territory. In one bombing raid Flight-Sergeant R. H. Middleton, of Yarra-bandai, New South Wales, a great-nephew of Australian explorer Hamilton Hume, won the V.C. Captain of a Stirling bomber, Middleton set out in November, 1942, to smash the Fiat works at Turin, then turning out tanks for the enemy. He had to burn more fuel than he could afford in getting there, then ran into heavy flak over the target. Shell splinters ripped through Middleton's right eye and into his legs and body. Middleton blacked out. The second pilot took over and the bombs sped down on target.

By then Middleton was conscious again. He ordered the second pilot back for first aid, and nursed his crippled plane over the Alps for the long flight home. He was in great pain and blood gushed from his wounds whenever he gave an order. He had constantly to wipe a hand across his face to keep the blood from his good eye. With his petrol running out as he crossed the British coast, Middleton ordered his crew to bale out to safety, then, rather than risk crashing on a sleeping house or village, turned back to meet certain death in the sea.

Another spectacular feat by Australian airmen came in February 1944, when Australian, New Zealand and British squadrons of Mosquito bombers were detailed to blast a hole in the wall of Amiens Jail and free 700 French Resistance fighters waiting trial or execution there. The attack took the Germans completely by surprise. The New Zealanders were first in. Each planted two 500 lb. bombs which ripped large holes in the outer wall of the prison. With split-second timing the Australians swept along next, to plant their bombs squarely on the end of the main cell block. They did so well that the British squadron was ordered back before reaching target. The prisoners were ready and before the German guards could rally they were scrambling over the rubble to freedom. Within minutes 258 had escaped and were scattering over the fields. Some were recaptured, tied to the bullet-pitted stakes in Execution Gully and shot. The main resistance leaders, however, got away.

Chapter 23

More giants of the air war

Two more Australian airmen won the V.C. in World War II. Both stood well over six feet; both flew unfalteringly at roof or tree-top level to smash the targets assigned to them. One, Hughie Idwal Edwards, rose to be Air-Commodore and A.D.C. to the Queen. He has the letters V.C., C.B., D.S.O., O.B.E., and D.F.C. after his name.

The other, Flight-Lieut. William Ellis Newton, V.C., was shot down off the coast of Salamaua and was murdered — beheaded by a Japanese with a Samurai sword. Newton had a premonition he would not return. "You won't be seeing me again," he told his mother on his last leave. He indicated a bottle of sherry on the mantelpiece. "Don't make a fuss," he said. "Just have a drink for me."

Hughie Edwards was born at Fremantle on August 1, 1914, his father, one of the dying race of grand old village blacksmiths. Hughie lost his first job as shipping clerk in the depression, after which he joined the Australian regular Army and in 1935 was one of 20 chosen from a thousand for cadetship in the R.A.A.F. In 1936, with clouds of war piling up over Europe, Hughie Edwards was given a short-term commission in the R.A.F. He nearly lost his life when a Blenheim he was flying crashed in England and for two years was in and out of hospital.

By then Hitler's war had erupted. Hughie Edwards was one of the Few, his job in light bombers being to smash Nazi shipping and strong points on occupied coasts. By 1941 Edwards was a Wing-Commander with a reputation for dare-devil hedge-hop attacks. Shy and unassuming, he was a strict, some said fierce, disciplinarian, with a liking for *"Gray's Elegy in a Country Churchyard"*. His men would follow into whatever hell he led them.

Edwards won his D.F.C. in June 1941, when he found a convoy of eight enemy ships moored off The Hague. As always, he flew in low. Through a hail of flak, he picked out a 4000-tonner, raked it with his forward machine-guns, and dropped his bombs at mast level. His aircraft reeled from the blast as debris shot into the air. He left the wreck sagging in the water.

In July 1941 Edwards was detailed to lead what is known as the High Tension attack on factories and docks in Bremen. With the port still burning from a night blitz, Edwards led 15 bombers on a daylight raid to spread more havoc and terror in docks and factories. Enemy ships sighted him and flashed his approach to shore stations. All ground and air defences were therefore on alert as Edwards led his formation across 50 miles of enemy territory. Three Blenheims were hit by flak and turned to stagger home.

Edwards and his squadron could not miss Bremen, for fire and smoke still rose from battered targets. They flew through fires at zero height, ducking under or bobbing over high-tension cables.

They had to dodge the cables of a formidable balloon barrage which would have ripped off their wings. Flak rose in a wall, but Edwards and his Blenheims flew on. They dropped their bombs on timber yards, factories, railway junctions and docks, the Blenheims lurching as debris shot 700 feet into the air. They were so low that they could see frantic Bremeners rush for cover under cars and behind buildings. With the bombs gone, they machine-gunned gun emplacements and raked barracks and railway depots from a height of only a few feet. Four of the Blenheims were shot down.

Edwards then marshalled the rest of his formation and led them safely through the balloon barrage and home. When they landed they found that every aircraft had been hit. For this deed of valour Edwards received the V.C. From then he was a leader of desperate ventures.

On December 6, 1942, Edwards was chosen to lead 94 bombers without fighter support against the Philips' radio factory at Eindhoven, Holland, which supplied one-third of all radio valves made under Nazi control. As usual Edwards took them at low level through rain squalls and cloud that reduced visibility to less than one mile. Three bombers were shot down, but the rest rained their high explosives on factories and workshops, smashing delicate equipment and destroying stocks. Australian Venturas then swept in with incendiaries. When they turned to fight their way out, the valve-testing shop and other buildings were burning wrecks. Thirteen aircraft were lost from that force. Many were damaged not just by gunfire but by seagulls and ducks, which crashed through the perspex windows of low-flying planes. Soon after this Edwards was promoted Group-Captain and put in command of a station where an Australian squadron was posted, but he still flew on bombing missions over Berlin.

$$* \qquad * \qquad * \qquad *$$

While Edwards was flying against targets in occupied Europe, 6ft. 3 in, 16-stone Big Bill Newton was smashing at the Japanese in New Guinea. Newton was born at St. Kilda, Melbourne, in 1919. His mates at Melbourne Grammar School knew him as a dashing cricketer and footballer. He joined the Air Force on the first day of war and was quickly noted as a born leader of men. In May 1942, he and the 22 Attack Squadron flying twin-engined Boston bombers were posted to New Guinea, where they battered Jap air-strips and dumps on the northern coast.

Newton won a reputation for unswerving daring and scorned evasive action. When he struck, he went straight for his target at tree-top level through heaviest anti-aircraft fire. The Japanese came to know and hate Newton's bomber as it screamed over the tree-tops to its mark. His admiring mates called him "The Fire Bug", because, whenever he flew, he left a fire raging in some enemy dump. Newton's luck began to run out early in 1943 when Japanese fire wrecked an engine after he had bombed the air-strip on Salamaua Isthmus. Newton kept control with difficulty and limped on one engine 160 miles before making a forced landing.

In March 1943, he set out to destroy buildings and dumps the enemy had established on the isthmus. The dumps included two 40,000 gallon fuel installations. Newton flew through half a mile of flak to bomb his target, his bombs starting a fire which other aircraft widened. As Newton turned away, shell splinters ripped through his fuselage and wings. Others pierced the petrol tanks and seriously damaged the main planes and engines. Again he had to fight for control. As he headed for base he saw that the fire he started was sending flames 1000 feet into the air and smoke to 8000 feet. Newton flew his wrecked plane 180 miles back to base and landed successfully despite a flat tyre.

He was now due for a rest after 52 sorties but would not take it. Back at Salamaua, he said, there was a single mysterious building flanked by A.-A. guns which just had to be smashed. Newton took off to smash it. Again he came in low through a hail of flak and saw his bombs wreck the building. Then a shell ripped into his plane which burst into flames. Newton turned to sea and made a forced landing half a mile from shore. His crew swam for it, Newton and Flt.-Sergeant John Lyon being the last to leave the burning aircraft.

When they reached the beach Newton and Lyon were captured by Lieutenant Komai and his squad of Japanese soldiers. Komai was impressed by the giant, easy-going Australian so devoid of fear and sent Newton and Lyon up the coast to Admiral Fujita's headquarters. Fujita had them grilled for information which they steadfastly refused to give, until the Admiral lost his temper and ordered them to be executed. They bayoneted Lyon on the spot, but Newton they sent back to Lieut. Komai with orders for him to execute the big Australian personally, according to the Samurai code. Newton was only 23 when he died. Even the Japanese who watched admired the courage and dignity with which he met death. Maybe his best epitaph is the concluding words of his V.C. citation: "The story of his brave deeds will become a legend and will be read with pride for years to come."

When the war ended, aviation in Australia faced a bright future. The day of the flimsy planes of the pioneers had passed. The fighters, the bombers, the Flying Fortresses, all had sparked new trends in design. Before the war, too, an R.A.F. engineering genius, Frank (now Sir Frank) Whittle, had invented the revolutionary new jet engine. Both sides in the conflict delayed too long to apply the engine to war, but it was advancing towards perfection when peace came. With the jet came the age of supersonic speeds. Soon Australia was to be less than 24 hours from Britain.

— Chapter 24 —————————————————

The Christchurch race

With the war nearing its end, it was abundantly clear that aviation had shed its growing pains and would emerge as a comfortable and safe form of transport, combined with hitherto incredible speed. No longer would it be an adventure to fly the Pacific or the Atlantic. During the war, Australian and British Air Force pilots ran a regular ferry service, flying bombing and transport planes and air crews from America and Canada to Britain, while the North Pacific route, pioneered by Kingsford Smith and Ulm and Smith and P. G. Taylor via Hawaii, was in constant use. In 1944, P. G. Taylor, now famous as Australia's greatest pathfinder and knighted for his service, surveyed a route from Mexico and in 1951 another to Chile. As jets came off the assembly line, it was plain that great air-liners would soon link Australia with Britain in less than a day and circle the globe at fabulous speeds and with the regularity of suburban trains.

It was also just a matter of time before some sporting or patriotic body came up with an idea for another great race in which the newest of the jet-driven monsters cleaving the air would clash in friendly, or not-so-friendly, rivalry. The first to seize on the idea were the citizens of the wealthy pastoral district of Canterbury, New Zealand, who thought such a race from Britain to their capital, Christchurch, would be an ideal way of telling the world of their centenary, due in 1953. They formed the Canterbury International Air Race Council and called for entries for first prizes of £10,000 in both speed and transport sections and total prize money of £29,000. The object of the race was to demonstrate to the world the shrinking of distances in point of time and give British manufacturers an opportunity of proving the excellence of their planes in open competition with those of other countries, in particular, it was hoped, America, over a gruelling course.

News of the race was hailed with enthusiasm. The Royal Aero Club in London agreed to take responsibility for the race from London to Basra, where the Canterbury Air Race Council would take over and manage the rest of the route to New Zealand. Tentative inquiries came from Britain, the United States, France, Holland, Denmark and Australia.

Despite the cost of the project, several private individuals planned to fly. Foremost among them were R.A.A.F wartime Squadron-Ldr. A. J. R. (Titus) Oates, and Flight-Lt. D. H. Swain. Oates had won the D.F.C. for attacks on Japanese shipping at Rabaul. He became a test pilot for De Havilland's, had luck in a lottery and owned a hotel at Campbelltown near Sydney, when the race was announced. Swain had also won the D.F.C. — for bombing raids over Germany. Both were brilliant airmen. The Australian Government promised them the use of an R.A.A.F. Mosquito for the race.

Interest flared higher when it was announced that the England-Australia woman pilot, Miss Freda Thompson, planned to captain an all-women's crew consisting of Mrs. Gertrude McKenzie, of Brighton, Vic., radio operator and relieving co-pilot, Miss Constance Jordan, of Rose Bay, Sydney, Australia's only woman aeronautical engineer as maintenance engineer and relief pilot, and Mrs. Evelyn McKoren, of Mount Gambier, relief pilot. The R.A.A.F. agreed to lend them the Duke of Gloucester's Avro Anson at a nominal charge. Investigation showed, however, that the Avro Anson possessed inadequate range, while the political situation along the route, particularly in Malaya, was, to say the least, uncertain. This tentative entry was cancelled. Mrs. Gertrude McKenzie and Miss Constance Jordan, however, still hoped to fly. Their chances seemed good when it was announced that a DC3 aircraft owned by Mrs. Diana Bixby, of California, was being entered in Mrs. McKenzie's name as the only Australian entry in the transport section. Mrs. Bixby would captain the crew while the second co-pilot would be Miss Betty Haas (an American) of Bixby Airlines. Mrs. McKenzie, it was stated, planned to go to California and help fly the plane to England via Iceland. Costs, however, were

Captain T.W. (Shorty) Shortridge, pilot of the ill-fated "Southern Cloud". Its loss remained a mystery for 27 years.

Tom Sonter, who found the wreck of the "Southern Cloud", high in the Snowy Mountains, 27 years after it was lost. His foot is on the centre engine of the plane.

Air Commodore A.H. Cobby, C.B.E., D.S.O., D.F.C. (two bars), G.M., one of Australia's first great airmen in war. His career included both World Wars.

high so the women pilots hoped to carry passengers to help with the finance. The following advertisement appeared in the New York Press: "Zoom to New Zealand. Fly as passengers with the fabulous female pilots entered in the London-New Zealand Air Race. Share thrills, glory, international acclaim and expenses. Thirty thousand miles of racing and a leisurely, fun-filled return."

Also a tentative entry was American Miss Jacqueline Cochran, wife of the President of an American aircraft corporation, who had been forced out of the Melbourne Centenary race with wing trouble.

In May, 1953, it was announced that there were 19 entries representing Britain, the United States, Holland, Denmark, Australia and New Zealand. The British, Australian, New Zealand and Danish air forces had entered and there would be 15 types of aircraft. As the weeks passed, however, the cancellations came. Costs were excessive. It was estimated that the Canberra jets entered in the race cost about £700,000 each and would eat up £8000 worth of petrol, not counting administrative and servicing expenses on the journey. Such costs and risks were beyond reach of all but wealthy nations. Besides, everyone conceded that the New British Valiant or one of the Canberra jets would win, though Squadron-Leader "Titus" Oates still believed he had a chance in his old-fashioned piston-engined Mosquito. The favourites in the handicap section were those persistent fellows the Dutch who, while protesting they were not racing at all, had won the handicap section in the Melbourne Centenary race and now declared they would carry in the Christchurch race 65 passengers, including specially-selected migrants who wanted to settle in the Canterbury district. Meanwhile, another Australian Flt.-Lt. J. L. Whiteman dropped out for financial reasons.

There was general regret when Britains Royal Air Force withdrew their glamour plane, the four-jet, 600 m.p.h. Valiant bomber from the race. It had not been completed in time for the necessary long-range test flights to be made. This left only six planes in the speed section, two Australian-built twin-jet Canberra bombers, entered by the Royal Australian Air Force, three British-built Canberra bombers, entered by the British Royal Air Force, and "Titus" Oates and D. H. Swain in their Mosquito. Three planes were entered in the handicap section, the Dutch K.L.M. Douglas with its 65 passengers, a BEA Viscount and a Royal New Zealand Air Force Hastings.

Titus Oates and Douglas Swain left Perth for Britain in their Mosquito on October 3, 1953, expecting to make the journey in three days and arrive two days before the start of the race. They had a fair flight to Cocos Island and were heading for Colombo in Ceylon at night when they ran into a fierce electrical storm high above the Indian Ocean. Lightning played round their aircraft. First the artificial horizon, then the airspeed indicator and altimeter went dead from lighting strikes. The electrical compass was useless; the radio receiver went dead. They were flying blind through torrential rain driven by a 100-knot gale. They could get neither above nor below the storm They just had to bash their way through. Soon they were hopelessly lost over a waste of water.

The first indication that they were in trouble came at 12.43 a.m. on October 4 when the Australian radio station at Port Hedland picked up an SOS from Oates asking radio stations for bearings to pin-point his position.

Soon after this, the radio receiver kicked into life on the lost aircraft and Oates found a flicker of response on the beam from Bangkok. He did not know if he had enough fuel, but it was his only lifeline, so he headed east for Bangkok. Any radio contact was better than none.

Port Hedland and Singapore continued to pick up distress signals. In one, Oates reported that he was trying to fly to Bangkok on the Bangkok radio beam. Then, at 4.40 a.m. came the ominous message that he was preparing to ditch the aircraft. Eleven minutes later he reported that he was approaching a coastline and would have to ditch into the sea within five minutes. He said he was hopelessly off course and did not know where he was. Then came the final message "I can see an island with a light and am making for it."

Meanwhile, a search was already under way. R.A.F. air-sea rescue flights took off from Singapore soon after the first distress signal was received. They concentrated on the Bay of Bengal. A DC6 left Cocos Island, while a Qantas Skymaster, three hours out of Cocos, was diverted over the search area.

Oates and Swain had been unlucky. Sixteen more minutes of gas and they would have made Bangkok. As it was, they ran out of fuel while approaching Bai-u-gan, an island in the Mergui Archipelago, off the Burmese Coast. Oates just had to ditch the craft and, as he pointed out later, Mosquitoes were not the sort of plane to ditch for light entertainment. "Normal landing speed is about 140 knots," he said, "and you need a landing run of about a mile and a half. Put down belly first into the sea, they usually porpoise, that is, they skip, dive deeply, then bounce back. The air screws chew through the cabin and amputate the pilot's legs. And the deceleration breaks his neck." Oates and Swain took all the seat cushions and packed them round the cockpit to protect themselves from the shock of deceleration when they ditched.

The Mosquito hit the water at something more than 140 m.p.h., and bounced. It skipped back to low-lying rain clouds, turned a complete somersault and belly-landed on a tidal mud flat a quarter of a mile off-shore. The shock ripped away the underbelly of the plane. Swain was thrown clear through the broken roof and flopped into the mud. Oates was trapped by the legs in the cockpit. Swain half swam, half crawled to the plane in time to save Oates from sinking in the mud which oozed into the cabin. They were still in a desperate plight when Burmese fishermen slithered over the mud, robbed them, rescued them and took them in their boat to a fishing village and then to Mergui where an English-speaking Indian doctor tended their injuries. R.A.F. planes later flew them to Butterworth and on to Ceylon where they got a lift home on the New Zealand Hastings freighter which had been forced out of the race by engine trouble.

Meanwhile, more than 30,000 people at London Airport saw the Duke of Gloucester flag away the competing planes on the 12,500 mile Canterbury Centenary Race from England to Christchurch, New Zealand. Though the racing section was now restricted to the two Australian Canberra twin-jet bombers entered by the Royal Australian Air Force and the three British Canberras entered by the R.A.F., the

inter-force rivalry was keen and the race was full of thrills.
Favourite was the brand-new British Canberra flown by
Wing-Cdr. L. M. Hodges, with the Australian Canberra,
piloted by Wing-Cdr. D. R. Cuming, Australia's No. 1 test
pilot, second in the betting.

It was a neck-and-neck race as far as Colombo, Ceylon.
First to touch down there was the favourite, Wing-Cdr.
Hodges, followed quickly by Wing-Cdr. Cuming and
Squadron-Ldr. Peter F. Raw, piloting the second Australian
Canberra. All five planes arrived within minutes of each
other, having completed the first half of the 12,500 mile
course in under 10½ hours. The Australian planes made up
for their small lag in time by speed of refuelling, which put
Raw in the air again one minute ahead of Hodges, with
Cuming lying a close third.

Bad luck lay in wait for Cuming. He was in a strong
position as his Canberra skimmed in to a good landing at
Cocos. Hardly had his wheels touched the strip than blue
smoke came from his tyres. Seconds later there was a loud
explosion and another cloud of blue smoke. The left tyre had
blown out. The aircraft, taxi-ing at 50 m.p.h., swayed.
Cuming said later that he used a short, sharp Anglo-Saxon
word as he held the bomber straight, then turned it off the
runway to keep the airstrip clear for other planes. Experts
among the onlookers praised his handling of the bomber.
"Any other aircraft bursting a tyre at that speed would have
been a write-off," they said. There was no spare tyre at
Cocos, so Cuming was out of the race. "It's no good being
annoyed," he said, "It's just one of those things."

Meanwhile, Hodges, in his brand-new Canberra, had
by-passed Cocos and raced direct from Colombo to Perth.
Despite this, Peter Raw, in the remaining Australian
Canberra, was the first to cross the West-Australian coast on
the leg which carried him to Woomera Rocket Range. At this
stage, fate again lay in wait for the leaders. Hodges was held
up at Perth by a burnt-out generator and a faulty fuel tank
cap. Raw, now tipped to win the race, found, while landing
at Woomera, that the cold air at 40,000 feet had frozen the
nose wheel of his 'plane. He could get the wheel only half
way down but, with great skill, landed the plane without

damage. Part of the air-speed measuring equipment broke as mechanics tried to force the wheel down. While they toiled to repair this, the other British Canberras, piloted by Flight.-Lt. R. Macalastair Furze and the rank outsider, Flight-Lt. Roland Burton, raced level. When Burton passed Melbourne, Raw, still grounded, was given no hope.

He was still in the race, however. One-and-a-quarter hours after he landed, he was in the air again, straining his plane to the limit in an all-out bid to catch the Englishmen. It was later stated that the three planes must have flashed over the Australian East Coast within minutes of each other, with Raw further north than the R.A.F. airmen. Crowds, waiting at Harewood 'drome Christchurch on that bleak dawn, tensed as they heard of the three planes racing all-out across the Tasman. They hoped, perhaps naturally, that the Australian would win and there was great excitement when a red port light was seen blinking through the half light four miles away. Cheers rose as the R.A.F. Canberra, piloted by Flt.-Lt. R. L. E. Burton with Flt.-Lt. T. H. Gannon, navigator, landed, having completed the course in 23 hours 51 minutes at an average flying speed of more than 540 m.p.h.

Second to land was the R.A.F. Canberra piloted by Flt.-Lt. R. Macalastair Furze, with Flt.-Lt. J. W. Harper as navigator. Hard after them came the Australian-built Canberra flown by Squadron-Ldr. Peter F. Raw with Flying-Officer F. N. Davies as co-pilot and Flt.-Lt. W. D. Kerr, navigator. As Raw had left London some minutes after Furze, he was second on adjusted time.

And what of the transport handicap section? It was won, of course, by the staid Dutch K.L.M. Douglas Air-liner which, commanded by Capt. Hans Kooper, flew sedately along its normal route via Rome, Bagdad, Karachi, Rangoon, Djakarta, Darwin and Brisbane with its cargo of specially-selected migrants, including a number of brides. Second came the British European Airways Viscount piloted by Capt. W. Baillie. The third entrant in this section, a Royal New Zealand Air Force Hastings withdrew at Negonbo, Ceylon, with engine trouble.

Since the London-Christchurch air race, air speeds have shot ahead. L. P. Twiss, of Britain, passed the 1000 m.p.h.

mark with 1132 miles an hour in 1956. J. W. Rodgers of the
United States passed 1500 m.p.h. in 1952, while America's
Lockheed won back the record from the Russians when their
new revolutionary YF12a recorded 2062 m.p.h. In the
experimental field, an American plane powered by rockets is
said to have reached nearly 4000 m.p.h.

In the domestic field, too, progress has been spectacular.
Where Ross and Keith Smith took 28 days to fly from Britain
to Australia, Qantas jets, flying to regular schedule, link
Sydney with London in 34 hours and bring migrants out on
charter flights in less than 24. Aviation and aviators have
come far since Colin Defries crashed his flimsy Wright biplane
on Sydney's Victoria Park racecourse in Australia's first
attempted powered flight in December 1909.

— Chapter 25 —————————————

The rest of their lives

Many of the aviation pioneers who feature in these pages had interests and adventures outside the scope of flying or made notable flights in other parts of the world. These notes will help to fill in the rest of their lives.

Dr. William Bland, the Sydney surgeon who invented the atmotic ship, something like a primitive Zeppelin driven by steam and studding sails, which he thought would carry passengers and freight from Sydney to London in four or five days, took a prominent part in the early turbulent history of New South Wales. After being transported for killing a purser in a duel, he was quickly pardoned and became Sydney's first, full-time private practitioner. He joined in the political sniping against Governor Lachlan Macquarie in 1818, lampooning the Governor in rhyme for putting his name on so many public buildings. He likened Macquarie to "that rude yonker, that with charcoal scrawls and marks his name upon the schoolhouse walls." For this Macquarie had him jailed for a year and fined £50.

Bland then joined W. C. Wentworth, the ardent patriot, in the first fight for self-government and was elected with Wentworth to represent the City of Sydney in the first partly elected Legislative Council. He was Secretary of the Australian Patriotic Association which fought for the rights of emancipists and also for the continuation of transportation in the interest of cheap labour. Bland took a prominent part in forming the Australian Medical Association and was a founder and treasurer of Sydney's free grammar school. He died in 1868.

*   *   *   *

Harry L'Estrange, whose balloon ascent in Melbourne in 1879 ended so disastrously, had already become famous for walking on a tightrope over Sydney's Middle Harbour. L'Estrange, who learned to walk the tightrope in his mother's back yard in Melbourne, thought he could get big money by excelling the mighty Blondin, who was said to have made nearly £50,000 from an Australian tour after he had walked across Niagara on a tightrope. Blondin performed in a tent in the Sydney Domain. L'Estrange decided to walk a tightrope across Middle Harbour. The rope would be 1400 feet long and 340 feet above the water, longer and higher than Blondin's Niagara rope.

L'Estrange launched his tightrope stunt with a great ballyhoo campaign. His agents whispered that it was suicide, so dangerous that the Government would ban it. Newspapers revealed that the rope between the two pylons on opposite sides of the harbour was in reality two ropes spliced together in the middle. The spliced section was thicker and sagged dangerously. It would probably send L'Estrange toppling to death into the harbour. If he did not fall, he would not be able to negotiate the last 100-yards uphill section and presumably would struggle there till he fell off from sheer exhaustion.

While the city seethed with rumours, L'Estrange chartered every free ferry on the harbour, 21 in all. Eight thousand people paid high prices to sail in them on March 31, 1877, to see L'Estrange tightrope over the harbour. L'Estrange had been very busy. His henchmen had roped off the best points of vantage and bluffed citizens into grandstand prices for them. He had talked publicans into running liquor booths on a share basis. Only when the refreshment stalls put up sold-out notices did L'Estrange mount his rope.

The crowds tensed as in yellow tunic and red turban with a 60 lb. balancing pole L'Estrange ran lightly along the rope at 80 steps a minute. All eyes were on the 20 feet spliced section in the middle where L'Estrange was expected to fall. Women shut their eyes, some are said to have swooned, as he tripped on reaching the spliced section. He recovered his balance and reached the middle. There he paused, lifted his left leg, waggled it and placed it against his right thigh. He knelt on the rope, waved gaily in all directions to the crowd. Then, as the rope swayed wildly, he reclined on his back as easily as a man reclining in a hammock, pulled out a telescope and gazed in all directions on the spectators.

Nerves tensed again as he rose to his feet and stepped lightly to where the rope rose abruptly to the pylon. This was where the experts said he'd be stumped. They were wrong. He climbed steadily. Crowds cheered, whistles tooted and bands blared as he reached the platform. Admirers chaired L'Estrange to a boat for a triumphant tour of his ferry fleet. Harry L'Estrange cleared £10,000 that day. He repeated the performance a few days later. Governor Sir Hercules Robinson and Lady Robinson went to see him, but the crowds stayed away. He had made the feat look too easy. They were willing to pay to see him fall off, but not to see him take a stroll across the harbour on a fine summer day.

L'Estrange was about to pull out when a body of sportsmen gave him a complimentary dinner. They presented him with a small purse of sovereigns, a tribute, they said, to a very courageous gentleman. They pleaded with him to risk death again to give another show. Thus flattered, he agreed. When he went to charter the ferries, he found that the body of sportsmen had leased the lot, so he did not make much from his last tight-rope walk over the harbour. It was then that L'Estrange turned to ballooning to be saved by a tall tree when his balloon burst in Melbourne.

*   *   *   *

Lawrence Hargrave who, according to American Octave Chanute deserved to be the first man to fly, was born at Greenwich, near London, on January 29, 1850. His father John Fletcher Hargrave was a barrister practising in London at the Equity Bar. His father's health, however, broke down and he was advised in 1856 to come to the warmer climate of New South Wales where he became a judge. Lawrence was educated in England till he was 16, then joined the rest of the family in Sydney.

Judge Hargrave wanted Lawrence to read for the bar but he failed to matriculate and, instead, entered the drawing office and later the workshops of the Australasian Steam Navigation Company. There, no doubt, he gained the knowledge that enabled him to build the steam, compressed-air and petrol engines he designed for his model planes.

Young Hargrave, however, had a strong sense of adventure. A few years after his arrival in Australia, he took time off to accompany a Queensland friend on a six-month trip to the islands of the Gulf of Carpentaria. His appetite for exploration thus whetted, he became, at 22, a member of the management committee for a group of hare-brained young adventurers who clubbed together to seek gold in the cannibal-infested jungles of New Guinea.

They bought the leaky, near-derelict, ex-clipper-turned-collier *Maria* for £300. The first captain walked off saying she was not fit to be taken outside Sydney Heads. The young adventurers made the chief officer captain and raked the bars round Sydney Rocks to find a man to take the mate's job. Only five of the 75 aboard knew how to work such a ship. The rest of the crew were enterprising young madmen answering the lure of New Guinea gold. They sailed on January 25, 1872, and quickly found that the *Maria's* timbers were unsound, her masts rotten and her chronometer unreliable. Water drained through the leaky decks and drenched the berths. The young adventurers cheerfully manned the pumps. Storms blew up. Heavy seas swept away the tiller, wheel and part of the port bulwarks.

Only then did the adventurers have misgivings. Twenty of them demanded to be put ashore. The disabled vessel was heading, they thought, towards Townsville when a change of wind drove her on Bramble Rock on the Great Barrier Reef. The captain and four seamen climbed into the only serviceable boat, took two others on board and deserted the rest on the pretext of seeking help. The others, thus abandoned, cut down masts and spars to make rafts. They were hardly ready when *Maria* sank till only her masts showed above water. Young Lawrence Hargrave was rescued from a mast top where he had taken refuge. Twenty-one were drowned. One boat and two rafts made shore where 14 more shipwrecked men were massacred by natives. Hargrave and the 39 other survivors were taken back to Sydney by Captain Moresby in the warship *Basilisk*.

This narrow escape did not daunt Hargrave. In 1875, as

engineer on the brig *Chevert* he sailed for New Guinea with Mr. (later Sir) William Macleay's expedition. Macleay's original aim was to collect zoological and other scientific specimens. Business circles, however, urged him to find a suitable river island where a trading post could be established and the colonisation of New Guinea begun. The expedition tried to penetrate the Katau River by steam launch and risked *Chevert* in dangerous and uncharted waters before abandoning the colonisation project.

When the expedition ended, Hargrave, displaying the persistence which marked his aeronautical experiments, returned to New Guinea with three other men to explore the Port Moresby hinterland. With the explorer, Luigi Maria D'Albertis, he explored nearly 500 miles up the Fly River. Adverse currents, shoals and sickness forced the explorers back. Hargrave returned to Sydney where, in 1877, he was elected to the Royal Society of New South Wales. He became assistant astronomical observer at Sydney Observatory and married Miss Margaret Preston Johnson who bore him five daughters and one son. Hargrave left this post in 1883 when he inherited a comfortable income of approximately £600 a year.

From then he devoted himself to his life work, the study of the problems barring the way to human flight in heavier-than-air powered machines. Besides his discoveries in the field of aeronautics, however, Hargrave had other discoveries and inventions to his credit. He experimented with wave-propelled vessels, shoes for walking on water, a one-wheeled gyroscopic car, a one-wheeled velocipede and a screw-driven airship. He prepared plans for improving and deepening the port of Sydney and expressed views concerning an Eastern Suburbs railway.

Hargrave, however, was slightly off-beam when he declared that markings on the rocks near his Woollahra, Sydney, home were the work not of aboriginal tribesmen but of Spanish seamen who landed there in 1595, nearly 200 years before Governor Phillip founded the first colony around Sydney Cove. Hargrave's theory was that Spanish Captain Lopez de Vega, whose ship *Santa Isabel* vanished in volcanic disturbances while exploring the Pacific in 1595, had actually

survived, made the first landfall on the east coast of Australia
and had landed near what later became known as Sydney
where his men filled in their time making carvings on the
rocks. Hargrave read a paper on this theory before the Royal
Society but little credence was given to it. Lawrence
Hargrave, who came close to being the first man to fly, died
on May 24, 1915.

*   *   *   *

George Augustine Taylor who, in 1909, became the first
Australian to be airborne — in a glider he himself designed
and built — was one of those dynamic, brilliant people who
popularised the scientific inventions which abounded at the
turn of the century. He came to the fore first as an artist and
cartoonist contributing, among other newspapers, to the
Sydney Bulletin and London *Punch.* Before he built his
glider, he had experimented with equally revolutionary
wireless telegraphy, succeeding among other things in sending
messages between speeding trains. His experiments towards
locating sound by wireless were useful in World War I when
submarines had to be tracked. He also did experimental work
in the sending of photographs and sketches by wireless.

Taylor was appointed to advise the Australian military
forces on wireless and aeronautics before the first World War.
It was due largely to his campaign that the Commonwealth
Government erected their first wireless station in Sydney in
1912, while, as chairman of a committee of experts, he
helped frame Australia's broadcasting regulations in 1923.
Taylor owned and edited several magazines and was an expert
also on town planning. He wrote several books, including two
volumes of popular verse. He died following a painful
accident in 1928.

George Augustine Taylor's elder brother, Vincent Patrick
Taylor, better known as Captain Penfold, was famous as an
aeronaut showman, touring parts of Australia and the United
States with a primitive airship, balloons and parachutes.

*   *   *   *

Harry Houdini, whom some, rejecting the claims of Colin Defries and Fred Custance, still believe made the first faultless flight in Australia, was better known as a master-magician extraordinary. Born in America in 1874, Houdini was famous the world over for his miraculous escapes from bonds and shackles. He thrilled crowds by diving manacled from bridges, freeing himself under water and swimming out, by emerging bleeding and torn from steel shackles and leather straight jackets, by staying for an hour and a half under water in a leaden coffin and by burrowing to safety through six feet of soil.

Billed as Handcuff King and Escape Artist, Houdini started his career in sideshows at fairs and circuses, sometimes giving 20 shows a day for 12 dollars a week. On the circus round between magician acts, he often doubled as the wild man from Java, while his wife when not appearing with Houdini as his vanishing lady, worked as a clown. He got his big break in 1898 when he defied the Chicago police to hold him. All America laughed when he broke out of the manacles and cell in which they placed him after searching him, stripping him naked and sealing his mouth with plaster. His reputation built up when he freed himself from Scotland Yard shackles with a single tap. Then, in 1900, at Dresden he developed his under-water escape.

Houdini saw his first aeroplane at Hamburg in 1909. He tried to buy it from the aviator who refused to sell. Within a week, however, Houdini had bought a Voisin. Two weeks later he smashed it up on his first attempt to fly. "Broke propeller all to hell," was the laconic entry in his diary. He brought the plane to Australia in January 1910, and, in March, made his first flight at Digger's Rest. Harry Houdini died of appendicitis in 1926.

*   *   *   *

Ray Parer, who with J. C. McIntosh was second to fly from England to Australia, completing the flight in eight months in an aerial jalopy in 1920, continued to fly in Australia. He bought a DH4 single-engined plane from Mildura dried-fruit magnate C. J. de Garis in which he won the first aerial Derby, staged in Victoria. He and his cousin Mark Parer then failed in a bid to be the first to fly around Australia, accidentally crashing their plane at Boulder near Kalgoorlie. When the Bulolo, New Guinea, gold rush flared, Parer hit on the idea of flying food, equipment and miners to the field to save the long mule journey. He formed Bulolo Aerial Transport Services, a self-supporting, unsubsidised New Guinea Air Line. To get a plane to his base at Lae, he flew it from Moresby over the Owen Stanleys, being the first to cross the high ranges.

When World War II broke, Parer volunteered for flying service but his health was not very good and he was rejected. He, therefore, took to the sea and was chief engineer of an American ketch when 30 Japanese fighters strafed it in Douglas Harbour, New Guinea. After the war he went pearling at Thursday Island.

*   *   *   *

Sir Alan Cobham was already hailed as a great aerial pathfinder when he flew to Australia and back to survey the route in 1926. Cobham, who was born in 1894, started first as a pupil farmer, found the land did not suit him and went to work in the city, planning a career in commerce. He went straight into the war in August 1914, and served in France till 1917 when he was commissioned in the Royal Flying Corps, later the R.A.F.

Flying was in his blood. On demobilisation he was one of the first to make a career of civil aviation, carrying 5000 passengers in 1919. He specialised then in aerial photography and, in 1921, joined the De Havilland Company, makers of record-breaking machines. In 1921, when long flights were rare, he flew 5000 miles around Europe, capping it a year later with an 8,000-mile-tour of Europe and North Africa. He also started a Spanish air line to Morocco.

In 1923, Cobham expanded his range. He followed a 12,000-mile tour of Europe, North Africa, Egypt and Palestine by flying from London to Rangoon and back. He won the King's Cup Air Race. After flying to Australia and back, he was pilot commander of a flying-boat expedition

which circumnavigated for the first time the whole continent
of Africa. He promoted the Through-Africa Air Route
scheme and, in 1931, undertook an Air Ministry survey flight
with a multi-engined seaplane up the Nile to the Congo.

Another major task awaited Alan Cobham. For many years
he had been studying the problem of refuelling in flight. In
1948, he introduced the Cobham system of refuelling in
flight into the United States Army Air Force, following it in
1951 with the drogue-and-probe system which was adopted
by the Royal Air Force and the American Navy and Army
Air Forces. Under this system, a probe from a fighter or
bomber makes contact with a hose from an air tanker by
means of an intricate system of electrohydraulics. Non-stop
world flights are now commonplace. One of Sir Alan's
companies bought Meteor fighters from the R.A.F. as they
became obsolete, installed magic boxes as pilots, and sold
them all over the world as target aircraft. Sir Alan still thinks
nostalgically of his flying days. "Flying is a luxury, nowadays
and no longer an adventure," he says. "In the old days the
man in the pilot's seat saw a grand panorama. Now he flies at
40,000 feet and sees nothing."

*   *   *   *

Clement J. de Garis, the Mildura dried fruit magnate, gave
an enormous boost to aviation in Australia by proving its
value in live-wire commerce and publicity. He and his pilot,
Lieut. Francis S. Briggs, set several records criss-crossing
Australia. Lieut. Briggs, a World War I pilot, often flew
Australian Prime Minister W. M. Hughes between London and
Paris. Briggs and De Garis made history in 1920 when they
made the first private aircraft flight from Perth to Sydney,
covering the 2,475 miles in 21½ flying hours. They were not
so lucky on another occasion when the authorities handling
the royal tour by the then Prince of Wales asked De Garis to
fly the Prince's mail from Perth to Sydney in two days. The
alternative was that the Prince, who was due to sail for Japan
in the Battleship *Renown,* would have to go without his mail
for some months.

Briggs and De Garis set out with the Royal Mail. On that

trip they had the only crash De Garis had in his life. The D.H. biplane collided with a tree just after taking off from a South Australian air field. De Garis and Briggs escaped with scratches. The accident did not stop the Prince's mail. De Garis organised alternative transport to catch the Prince before he embarked at Sydney. As a token of thanks, the Prince gave Mr. De Garis an inscribed cigarette case. The log-books of Briggs and De Garis tell of flights in dust storms, forced landings in outback paddocks and of the pilot feeling his way across deserts without the aid of radio or modern navigation devices.

By live-wire organisation and superb propaganda De Garis made the South Australian dried fruit industry and brought it back to prosperity when it was faced with ruin. At 26, he had his own fruit-packing shed in Mildura. He staggered the neighbourhood by buying the derelict Pyap Village estate, reviving dying vines and modernising the picking, drying and packing of fruit on an assembly line basis. He built a model village, gave the settlers a school, library, billiard-room and a three-guinea bonus for every baby born on the estate. When the 1919 crisis came and strikes tied up dried fruit waiting for shipment on the docks, De Garis launched a nation-wide publicity campaign that soon had all Australia eating Mildura currants and sultanas. To help this campaign, he bought the planes in which he and Briggs wrote the name Mildura across the skies in record flights. By such means De Garis saved the industry from ruin and gave it lasting prosperity. He died tragically in 1926.

*   *   *   *

Frank Hurley, who explored parts of New Guinea by air in the flying crate days, lived a life of almost constant adventure. Beginning with a 15-shilling box camera he became Australia's No. 1 photographer. In that capacity, he made five trips to the Antarctic, led expeditions to the Northern Territory as well as New Guinea, was official photographer and film cameraman in two world wars and lived in Bedouins' tents and kings' palaces.

Hurley made his first trip to the Antarctic as official

Mr. C.A. Butler, an Australian who saw the great possibilities of commercial aviation and formed his own company to carry mails, passengers and freight.

Flt.-Lt. Roland L. Burton, pilot (left), and Flt.-Lt. T.H. Gannon, navigator, after winning the London–Christchurch Air Race in 1953.

photographer with Dr. (later Sir) Douglas Mawson who, after exploring with Shackleton in 1907-8, organised the first Australian Antarctic expedition in 1911. With Eric Webb, magnetician, and Robert A. Barge, astronomer, Hurley set out for the south magnetic pole. Pulling a sledge loaded with 790 lb. of food and equipment, the three men had to turn back when within 20 miles of their objective through lack of food. On this trip, they trudged 305 miles through sub-zero temperatures and were out two months.

On returning to Australia, Hurley was filming aborigines near the Gulf of Carpentaria when he received an invitation from Sir Ernest Shackleton to join his 1914 Antarctic expedition. Through no fault of Shackleton, the expedition ship, *Endurance,* was frozen in for ten months before pack-ice crushed it. The 28 members of the ship's company transferred to an ice floe with all the boats and equipment they could save. They drifted for six months on the ice and when the floe began to break took to the boats in which, after a few weeks sailing, they reached Elephant Island, 600 miles south of Cape Horn.

Shackleton and five companions then made the famous 800-mile journey through wild seas in a 21-foot open boat to the South Georgia Island whaling stations for help. In the meanwhile, Hurley and the rest of the party lived for five long months on raw penguin and seal meat and drank melted ice. Instead of tobacco, they smoked penguin feathers, seal hair and the grass padding from their polar boots.

Hurley and his comrades returned to London to find the war had been on two years. He became official war photographer with the A.I.F., had his baptism of fire at Hill 60 and was sent to the Middle East where he filmed the victorious compaigns of the Australian Light Horse. After the war, he explored New Guinea and made two more Antarctic expeditions with Sir Douglas Mawson in *Discovery*. He was one of the party which hoisted the Australian flag in Antarctica, taking possession of all the territories explored by Britishers there since the time of Captain Cook. In 1934, Frank Hurley set out with two companions to fly from Australia to England. The plane crashed in taking off from Athens.

In the Second World War, Frank Hurley was again an official photographer. For nearly three years, he filmed the A.I.F.'s battles in Tobruk and other parts of the Western Desert. He was sent to Persia to make films of the supply route to Russia and was an official photographer at the Teheran Conference where Churchill, Stalin and Roosevelt met to discuss the war. Hurley took pictures of the Shah and his first wife, Queen Fawzia, and of the then nine-year-old King Feisal II of Irak who 17 years later was shot by rebels. Hurley wrote several books, produced a superb set of photographic books and was working on another when he died, in January 1962, at the age of 71.

*   *   *   *

And what of the Mollisons, the redoubtable Jim and his wife, the former Amy Johnson. Both continued to flash across the headlines after their record-breaking flights between England and Australia. Amy, awarded a C.B.E., made a record-setting flight across Siberia to Tokyo and back and was on a leisurely sea trip recuperating from appendicitis when she met Jim Mollison at Cape Town, where he had just landed after breaking the England-Cape Town record. A few months later they married.

Records followed thick and fast for Jim Mollison — first solo westward flight over the North Atlantic, first westward flight over the South Atlantic, with Mollison a tiny speck over thousands of miles of ocean keeping going on a packet of chicken sandwiches, sticks of barley sugar and a thermos of coffee. In far-from-wifely fashion, Amy robbed him of his Cape Town record, setting new times for the flight there and back. Jim, following the previously unused eastern route, won the record back.

Jim and Amy then teamed to make record flights together. Those who knew them chuckled. They could not see such a partnership working. Both were individualists. Both would want to be the boss. There would be bickering over whose turn it was to fly. Perhaps that was why their joint flights were not a great success. Together they attempted the first non-stop flight from England to New York via Newfoundland, but failed for lack of petrol when 60 miles from their objective. They came down in a marsh. Their plane overturned. Jim was trapped unconscious in the wreckage and was saved by Amy who held his head above water till help came. After their failure in the Melbourne Centenary race, in which they smashed all records to Bagdad and India, both Jim and Amy returned to solo flying. Jim flew from Newfoundland to London. Amy set new records for London-Cape Town and back.

With such persistent rivalry few were surprised when Amy announced they were separating amicably. Besides her C.B.E., Amy Mollison held the President's Gold Medal of the Society of Engineers, the Egyptian gold medal for valour, the Women's Trophy of the International League of Aviators, the Segrave Trophy, the Gold Medal of Honour of the League of Youth, and the Gold Medal of the Royal Aero Club. The Gold Cup bought for her by the children of Sydney is now offered annually at her home town, Hull, for the most courageous juvenile deed of the year.

Apart from his flying, Jim Mollison was a most boisterous personality. Born in Scotland in 1905, he was one of the young dare-devils who learned to fly in the tough school of short-term commissions in the R.A.F. In 1928 he drifted to Australia and had been flying instructor at Adelaide Aero Club when Charles Kingsford Smith snapped him up as one of his A.N.A. crack pilots. Mollison was short, slim, boyish. His long hair lopped over his forehead; his voice was an Oxford bleat, so exaggerated that once when he asked for a bear, meaning *beer,* the barmaid snapped: "What will it be, brown, black or grizzly?" A six-foot squatter he encountered in a Sydney night club regretted bitterly his remarks on what he considered Mollison's "cissy" appearance. Mollison invited him outside and knocked him flat, proving he still had the punch that made him a top-flight lightweight in the R.A.F.

Between smashing records, Mollison lived hectically. He quickly made his mark in Australia as a stunt pilot. Many Sydney folk still remember the hair-raising dog fight he "fought" over Mascot with Travis Shortridge, another A.N.A. pilot. They whirled so low that spectators shrank in alarm. Only Horrie Miller, winner of the Sydney-Perth air race, ever bested Mollison in such a battle. Mollison buzzed Miller and a passenger at Adelaide. Both were in Moths. Miller landed, off-loaded his passenger and took off in a flaming rage. He buzzed Mollison so furiously and deftly that he forced Mollison to land.

Mollison's dislike of C. W. A. Scott helped him on the road to record-breaking. Scott, who flew for Qantas, had the same background, the same tough training, the same skill with the fists. They hardly bothered to conceal their hostility. It was to cut Scott down a bit that Mollison set out to break Scott's record for the Australia-England flight.

On parting from Amy, Mollison reverted to the out-and-out playboy. Women found him irresistible. He painted the French Riviera red, then switched to Le Touquet, the fashionable resort on the French Channel coast. After drinking one night with the Maharajah of Kapurthala and Richard Fairey, Mollison dashed out of the casino bar to his plane. He taxied madly about the 'drome, took off, buzzed a cafe terrace so closely that the 200-odd drinkers thought their last moment had come, then looped and spun a few feet above the beach, scattering the crowds in alarm. Amy had no difficulty in getting a divorce.

Mollison then married Mrs. Phyllis Hussey, a Jamaican plantation heiress with £ 20,000 a year. He turned over a new leaf, joined the Oxford Group Movement which he described as the most practical form of Christianity. His second marriage lasted 10 years. He then married Dutch-born Mrs Mary Kamphius. Somewhat tardily he was awarded the M.B.E. By then Mollison's hectic life was catching up with him. The one-time athlete and air hero was crippled by arthritis and could only walk stiff legged. He bought an unlicensed hotel in a London suburb where he talked over the golden days of flying with his friends. He died in November, 1959.

*   *   *   *

Tasmanian-born Harold Gatty who made the record round-the-world flight with Wiley Post in 1931 had a distinguished career both in American and in Fiji. Like

Kingsford Smith and Ulm, Gatty dreamed of a trans-Pacific service taking in America, New Zealand and Australia. His plan was much the same as Ulm's — a string of island bases with airstrips where land planes could alight. He hoped to organise such a company himself but the giant Pan-American Airways stepped in. Gatty, however, had some consolation when they appointed him to survey such an island route for them and negotiate for landing rights. Australia was antagonistic to the plans but New Zealand, after much argument, agreed. P.A.A. had inaugurated flights to New Zealand when World War II flared. The route was useful to the Allies.

During the war, Gatty was director of Air Transport for the Allied forces first in Java then at General MacArthur's headquarters in Australia and in New Guinea. Indirectly Gatty saved the lives of many torpedoed men at sea. For years he had studied the navigation methods of the Polynesians and how to survive when adrift on the ocean. He wrote a book *The Raft Story* and, when he could not find a publisher, bought control of a small publishing house and published it himself. As the war ground on, the U.S. Government put a copy of Gatty's book in the pack of every serviceman who might find himself alone on the Pacific.

After the war Gatty settled in Fiji. He had his own instrument-making factory, tried his hand at tuna canning which proved an expensive failure, and, returning to aviation, formed Fiji Airways to provide a link between the international airport at Nadi and Suva and linking Suva with other important centres. He then acquired an island, complete with bungalow and coconut plantation in the ro-mantic Lau group, two days sailing from Suva. He continued his study of ocean currents and the migratory habits of birds and even tried to solve the mystery of the October-November rising of the Palolo coral worm. He published another book, *Nature is Your Guide,* also aimed at teaching lost men how to survive. Harold Gatty served two terms as member of the Legislative Council in Suva. He died of a heart attack in 1957, aged 54.

*   *   *   *

Wiley Post, whom Gatty navigated on his round-the-world flight, was a Texas farm boy who was so engine-mad that he was mechanical handyman for the whole neighbourhood when he was 13. He wanted to be an airman in World War I but they would not have him. Instead, he went into signals. On discharge he became a roughneck on the Oklahoma oil fields, but threw this up to become a parachute jumper in a flying circus. For two years he thrilled thousands with spectacular jumps. He made his first flight in a circus plane, bungling it so badly that they urged him never to fly again. When the circus failed Post went back to the oil fields. The first day he was watching a roughneck drive an iron bolt into position with a heavy sledge when a splinter from the sledge flew into his left eye and destroyed the sight.

With 1800 dollars compensation for his lost eye, Post bought his first plane, a ramshackle crate with which he earned a precarious living. In 1928, he became personal pilot to a wealthy oil man, P. C. Hall, who two years later authorised him to buy a new Lockheed Vega which Post christened *Winnie Mae*. He gave the plane its first racing test in the Chicago National Air Race which he won. He then flew round the world with Gatty, following the 15,500 mile northern route, and, helped by a robot pilot, repeated the feat solo two years later, cutting a day off his previous record. Post then attacked the altitude record, reaching more than 49,000 feet, a record which was not recognised because of barograph trouble.

By then *Winnie Mae* was worn out. Post bought a Lockheed Orion seaplane and invited his old friend, Will Rogers, cowboy comedian and homespun philosopher of the Ziegfeld Follies, to accompany him on a leisurely tour of the world in the new aircraft. They ran into fog while flying from Fairbanks to Point Barrow in Alaska. Post alighted on a shallow lake to ask a party of Eskimoes the way to Point Barrow. He was taking off again when the engine failed, the right wing dipped and the aircraft crashed into the lake bank. Both men were killed, a tragedy that sent the whole American nation into mourning.

*   *   *   *

Sir Patrick Gordon Taylor, first, with Kingsford Smith, to fly the Pacific from Australia to the United States and who pioneered the South Pacific and Indian Ocean routes, was an "elder statesman" of Australian aviation till his death in Honolulu in December 1966.

*   *   *   *

Captain W. N. Lancaster, who flew from London to Australia in 1927-8 with Mrs. Keith Miller, the first woman to make the journey by air, later bought the *Southern Cross Minor* the single engined Avro Avian formerly owned by Kingsford Smith. He set out in the plane in April 1933, in a bid to break Amy Johnson's London-Cape Town solo record. Ignoring all warnings, he flew into the Sahara against a head wind and vanished. Twenty-nine years later, in 1962, a French camel patrol found the *Southern Cross Minor* in the desert with the mummified body of Captain Lancaster beside it. He had died of thirst.

Index

A